Texas Bobwhites

ELLEN AND EDWARD RANDALL SERIES

Texas Bobwhites

A Guide to Their Foods and Habitat Management

JON A. LARSON, TIMOTHY E. FULBRIGHT,
LEONARD A. BRENNAN, FIDEL HERNÁNDEZ,
AND FRED C. BRYANT

 University of Texas Press, Austin

The publication of this book was supported in part by the Caesar Kleberg Wildlife Research Institute at Texas A&M University–Kingsville.

This volume has been printed from materials furnished by the authors, who assume full responsibility for its contents.

All images by Jon A. Larson unless otherwise noted.

LIBRARY OF CONGRESS CATALOGING-IN-PUBLICATION DATA

Texas bobwhites : a guide to their foods and habitat management / Jon A. Larson . . . [et al.] — 1st ed.
      p.   cm.
   Includes bibliographical references and index.
   ISBN 978-0-292-72369-6 (cloth : alk. paper)
   ISBN 978-0-292-72278-1 (pbk. : alk. paper)
   1. Northern bobwhite—Food—Texas. 2. Seeds—Texas—Identification.
3. Northern bobwhite—Habitat—Conservation—Texas. I. Larson, Jon A. (Jon Andrew), 1979–
   QL696.G259T486 2010
   639.9′78627309764—dc22   2009051463

This book is dedicated to

Ed and Ellen Randall,

Morton and Bobbie Sue Cohn,

and

the South Texas Chapter of Quail Unlimited

for their

generous contributions and support

that made this book possible.

Contents

Preface

Northern bobwhites are one of the most popular game birds in the United States. More than 100,000 hunters pursue bobwhites each year in Texas. Scientists at the Caesar Kleberg Wildlife Research Institute frequently receive packets of seeds that hunters have removed from bobwhite crops along with a query as to what plant species produced the seeds. Interest of hunters and the general public in the food habits and ecology of bobwhites is something that we would like to encourage with this book. Conservation of both wildlife and the habitat they depend on for survival will be enhanced by greater understanding by the public of the ecology of wildlife.

Maintaining quality northern bobwhite habitat is economically important to Texas landowners and to the Texas economy in general. Texas ranchers with good bobwhite habitat often generate a greater proportion of their income from trespass fees paid by quail hunters than from livestock production. At the time of this writing, landowners commonly gross $10–12 per acre from quail hunting, with even greater amounts for quality habitat with large numbers of birds. In addition, hunters in Texas spend around $1.5 billion annually on trip-related expenses, equipment, and other items associated with hunting. Thus, availability of good bobwhite hunting in Texas benefits a wide variety of business interests in Texas.

The objective of this field guide is to provide a pictorial guide to the seeds commonly eaten by northern bobwhites in Texas. The authors hope that interest in what constitutes good bobwhite habitat among hunters and nature enthusiasts will be enhanced by this guide to the identification of seeds eaten by quail.

The task of arriving at the species to include in this field guide was a difficult one. Bobwhites eat the seeds of hundreds of different plant species, and ranking the importance of each species is somewhat subjective. We reviewed scientific journal articles, popular articles, and other literature describing food habits of bobwhites in Texas. We used the common names in the USDA Natural Resources Conservation Service plants database (http://plants.usda.gov/) in the text; we also provided alternative common names of plants that are widely used. Each species of seeds listed in studies of food habits was ranked based on volume or mass (weight) of the seed and the frequency at which the seed occurred in bobwhite diets. Many species listed in the field guide may be locally or seasonally important, but overall, we feel that we have a representative list of seeds that are important to bobwhites in Texas. Because the seeds were photographed at high magnification, and because many of them have detailed features, we recommend using a magnifying glass to assist in proper identification in the field. A glossary at the end of the book provides definitions of technical words used.

Bobwhite populations are declining by 3–4% per year in Texas. This decline is, at least in part, a result of continuing habitat loss and degradation of remaining habitat. A second objective of this field guide is to detail management actions to conserve and improve northern bobwhite habitat. Proper management of bobwhite habitat and restoration of bobwhite habitat where it has been destroyed are critical to the survival of the species. Habitat restoration, in particular, will become increasingly important to conservation of northern bobwhites in the future. Planting of, and invasion by, exotic grasses looms large as one of the major factors causing degradation of bobwhite habitat, and it stands as a major obstacle to restoring northern bobwhite habitat. We have included a chapter detailing this threat and the state-of-the-art management approaches to dealing with it.

Acknowledgments

Many people provided valuable assistance in the production of this field guide. We extend our special thanks to Morton Cohn, Ellen Randall, the South Texas Chapter of Quail Unlimited, the South Texas Quail Associates Program, the Richard M. Kleberg, Jr. Center for Quail Research, and the Caesar Kleberg Wildlife Research Institute for financial support. We thank Dr. Robert I. Lonard for verifying identification of voucher specimens of plants appearing in photographs in the book. We also thank anonymous reviewers who made helpful suggestions. We thank Charles Wissinger, Chair of the Department of Art at Texas A&M University–Kingsville, for providing line drawings. We also thank Hugh Lieck of Event Photography in Kingsville, Texas; Doug Smith, Campus Photographer, Texas A&M University–Kingsville; and Steve Bentsen for their advice and assistance on photography. We thank Clayton Wolter of the John G. and Marie Stella Kenedy Memorial Foundation and Donald C. "Chip" Ruthven III of the Texas Parks and Wildlife Department for their assistance in fieldwork. Thanks also to Eric Grahmann, Luke Garver, and the Faith Ranch for assistance in fieldwork. We thank Florence Oxley, Michael Eason, Damon Waitt, and Joseph Marcus of the Ladybird Johnson Wildflower Center for providing photos of *Desmanthus illinoensis, Galactia heterophylla, Lespedeza virginica, Morella cerifera, Panicum capillare, Rhynchosia latifolia, Sideroxylon lanuginosum,* and *Zanthoxylum clava-herculis* and for other assistance. Forrest Smith provided a photograph of *Urochola texana.* Marcie O'Connor provided a photograph of *Paspalum setaceum.* Thanks to Karen Tenaglia for providing images of *Lespedeza virginica, Scleria ciliata, Galactia regularis,* and *Lespedeza stipulacea.* Steve Bentsen provided a bobwhite photo for the frontispiece. Jim Muir and Randy Bow provided seedlings of *Strophostyles leiosperma* for photography. We thank Rachel Barlow of Texas State University for providing seeds of *Chamaecrista nictitans* and *Urochloa fusca* for photography. Imagery for plant range maps was provided courtesy of the Texas Commission on Environmental Quality and downloaded from http://www.glo.state.tx.us/gisdata/metadata/counties.htm. Thanks to Tony Falk for assisting in various phases of manuscript preparation. Also, special thanks to Paula Maywald and Forrest Smith of South Texas Natives for providing seeds, assisting in plant identification, and offering suggestions for this field guide.

Texas Bobwhites

CHAPTER ONE

Introduction

Why This Book Is Needed and Why It Is Important

Each fall and winter, sometime between the end of October and the end of February, nearly 100,000 Texans take to the field with shotguns in hand to pursue the bobwhite quail. Quail hunting in Texas, as any quail hunter will tell you, is serious business. Aldo Leopold stated, more than seven decades ago, that quail are grand-opera game. Texas is arguably the last remaining stage where sufficient habitat remains to provide quail-hunting opportunities on a grand scale. While good quail hunting remains in surrounding areas of Oklahoma, as well as parts of the Midwest and on private hunting properties in the Southeast, none of these areas comes close to providing the millions of acres of quail habitat and hunting opportunities that remain in Texas.

A great deal of planning, organization, and management underlie the factors that result in a successful quail hunt. At a minimum, hunting dogs must be trained, habitat must be managed, and shotguns must be cleaned. And this is before any quail hunting takes place. When you add up the costs of owning or leasing land, managing that land to produce an annual crop of quail, and the rest of the resources needed to support an operation, such as a hunting camp, vehicles, horses, and so on, quail hunting in Texas is an enterprise that pumps hundreds of millions of dollars into the economy of our state.

When quail end up in the quail hunter's bag, examination of the birds can tell the hunter a great deal about what they were doing earlier that day. This is because quail store the foods that they eat in a small but expandable sack called a "crop," located in the upper part of their chest cavity. During the fall and winter, most of the foods eaten by quail are seeds, and these seeds are stored in each bird's crop, usually twice a day—once in the morning after they leave their night roost and once in the afternoon or evening before they return to their roost. By quickly accumulating a large number of seeds in their crop during a foraging bout, the birds can then return to the safety of denser cover and process or digest their food away from view of the numerous predators who want to make a quail one of their daily meals.

Removing a quail crop and opening it to inspect its contents reveals a window into the life of that bird because it provides a direct record of what it had eaten earlier that day. However, the seeds of various plants eaten by quail, to the untrained eye, can be mysterious at best, or completely incomprehensible at worst. This is because the things—mostly seeds—that quail eat are usually totally unknown to humans.

This is why this book is needed, and this is why it is important. Any quail hunter in Texas can use this book as a field guide to identify the seeds that quail have eaten and then make the modest leap in logic toward knowing the plants that have provided these seeds.

By allowing quail to tell us what they have been eating, this information allows us, in various ways, to understand the components of their habitat that provides the foods that sustain them through the winter months. What quail eat changes with the seasons, of course. However, by giving the quail hunter a window into the winter diet of their quarry, which is what this book does, the hunters not only gain increased knowledge and perspective of these animals, they also gain knowledge of the plants that are important to sustaining them.

A final key point to realize is that the habitat that produces quail also supports scores of other species of wildlife in the rangelands and woodlands of Texas. By knowing what quail have eaten, people can then make an important and educated step toward conserving the habitat that sustains not only quail, but also many other species of wildlife that are a key part of the culture and heritage of all Texans.

CHAPTER 2

Food Habits and Nutrition of the Northern Bobwhite

The needs of wildlife can be summarized into four basic requirements: food, water, shelter, and space. Of these, food generally is the requirement that garners the most attention from land managers, particularly when involving game species. What are the preferred food items? What is their diet composition, and how does it change seasonally? How can management create habitat conditions with an abundance of food? Do practices such as supplemental feeding increase survival, enhance reproduction, or increase density? These questions represent common queries in the management of any game species, particularly the northern bobwhite.

Fortunately, the northern bobwhite represents one of the most intensively and extensively studied wildlife species. Food habits and nutrition of bobwhites were among the first topics to be addressed by early researchers. In fact, some of the first published studies on bobwhites, appearing in the early 1900s, focused on their food habits. This interest in general bobwhite ecology and food habits continued during the twentieth century through the work of people like Herbert Stoddard, Walter Rosene, and Val Lehmann. As knowledge accumulated on dietary needs and preferences of bobwhites, the research focus began to shift during the latter part of the century to more management-oriented issues such as supplemental feeding. Scientists and managers therefore enjoy an extensive knowledge base from which to answer nutritional queries and on which to base management decisions.

Herein we provide a concise review of the research on the food habits and nutrition of bobwhites. We do not discuss directly the value, or lack thereof, of management issues such as supplemental feeding or food plots. We do, however, discuss research on these topics tangentially as it relates to nutritional ecology of bobwhites.

Food Habits

Bobwhites are granivorous birds. About 60–80% of their diet is composed of seeds, primarily of grasses and forbs, with a lesser reliance on seeds from woody plants. They prefer seeds with a hard seed coat versus seeds that lack one. Thus, the seeds from plants such as ragweed, partridge pea, and brownseed paspalum represent more important food sources for bobwhites than do the seeds of plants such as little bluestem, hooded windmill grass, or hairy grama.

The remainder of the bobwhite diet consists of green vegetation, mast (i.e., fruits of trees and shrubs), and insects. Generally speaking, their diet consists of about 70% seeds and mast, 12% insects, 10% green vegetation, and 8% miscellaneous items. However, the relative importance of each of these components varies by season (figure 2.1). During the fall and

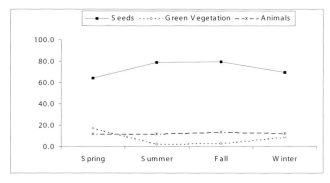

FIGURE 2.1. Seasonal composition of northern bobwhite diets throughout a year. Data represent mean values from Lehmann (1984) and Wood et al. (1986).

winter, seeds comprise about 70% of the diet, but as spring approaches and vegetation begins to grow, bobwhites switch from a diet primarily of seeds to one consisting of green shoots and insects. The dietary shift to greater insect use from winter to spring to summer can exhibit substantial increases (e.g., from 7% to 26%). Insects become increasingly important in the bobwhite diet during spring and summer because of nutritional demands associated with reproduction and growth of chicks. Insects are high in protein and represent an important source of moisture to breeding bobwhites during the hotter, drier summer months.

Given the overall predominance of seeds in the bobwhite diet, efforts have been made to categorize or rank bobwhite seeds by order of importance. In the 1950s, Verne Davison categorized 138 plant species as either "choice" or "inferior" bobwhite foods and referenced 378 species as "unimportant." Ranking bobwhite foods according to importance, however, is a rather difficult task, considering that bobwhites consume seeds and parts of more than 1,000 plants. In addition, nutritive value of seeds depends on several nutritive factors such as energy, protein, fiber, and digestibility and non-nutritive factors such as size and availability. We do not attempt to provide any type of ranking of seeds eaten by bobwhites, but rather we discuss the relative use of seeds based on percentage composition of crop contents.

Seeds of forbs commonly found in bobwhite crops include cuman ragweed, croton, and legumes such as partridge pea, snoutbean, and bundleflower. Staple grasses include panicums, paspalums, and bristlegrasses. In southern Texas, Val Lehmann reported that croton and panic grass comprised about 20% and 30%, respectively, of the bobwhite diet based on 565 crops. In the Rolling Plains of Texas, A. S. Jackson documented that seven forb species, which included croton, cuman (western) ragweed, and whitemouth dayflower, contributed 40% of the total volume of food in 963 crops analyzed. Seeds of paspalums comprised another 20% of the food consumed. Seeds from a relatively small group of grasses and forbs therefore comprise a large part of the bobwhite diet. Seeds and mast of shrubs are common food items in bobwhite crops, but the relative use of seeds from woody plants is less compared to that of grasses and forbs. Commonly eaten seeds of woody plants include gum bully, honey mesquite, hackberry, and algerita.

Bobwhites exhibit great diversity in their consumption of insects; however, grasshoppers often are the most common insects found in bobwhite crops. Other common insects include beetles, crickets, ants, termites, and spiders. The predominance of a specific category of insects in bobwhite crops can be influenced by its availability. A. S. Jackson reported that the crop of a bobwhite harvested in the Rolling Plains contained 136 winged ants. In southwest Texas, a study conducted on bobwhite food habits during a drought reported that desert termites occurred in more than half (52–57%) of all bobwhite crops analyzed. One quail crop contained 200 desert termites! The authors hypothesized that desert termites increased during drought because of favorable conditions for caste building, and that this increased availability likely contributed to their high occurrence in the bobwhite diet.

Bobwhites tend to show slightly different dietary preferences than their sympatric companion, scaled quail. Although both species use a wide variety of seeds, seeds of forbs and woody plants appear to be relatively more important to scaled quail than to bobwhites. As noted earlier, bobwhites generally consume more grass seeds. Bobwhites also appear to consume a lower diversity of foods (average of eight food items per crop) compared to scaled quail (15 food items per crop). However, bobwhites generally consume more animal matter (36%) than scaled quail (10%). Despite these slight dietary differences, bobwhites can experience a low to moderately high degree of dietary overlap with scaled quail, depending on the season and locality.

It is important to understand that the dietary items, composition, shifts, and species comparisons described above represent generalizations. Precise makeup of the bobwhite diet will vary by locality and from year to year. For example, studies in the Rio Grande Plains of Texas have documented that green vegetation can comprise as little as 10% (Kleberg County) or as much as 72% (Maverick County) of the bobwhite diet during fall–winter. In addition, consumption of insects may exhibit a distinctive peak of use such as during summer (Maverick County) or may exhibit an increasing trend from spring (8%) through summer (14%) into fall (18%) and winter (23%) (Kleberg County). Thus, although bobwhites are primarily granivores with a peak insect consumption during spring–summer, the bobwhite diet is quite variable across space and time. Food availability, weather, soils, vegetation, and numerous other factors influence its composition at any point in time.

Nutritional Requirements

Much of what is known about bobwhite nutrition can be attributed to the work of Ralph B. Nestler and Robert J. Robel. During the 1940s, Nestler began a series of experiments to determine the nutrients necessary for growing, breeding, and non-breeding bobwhites. His research focused on several macronutrients and minerals, including protein, calcium, phosphorous, and vitamin A. A few decades later, Robel continued this nutritional research by determining the energetic requirements of breeding and non-breeding bobwhites. A unique aspect of Robel's research was that it also evaluated how temperature affected the bioenergetics of bobwhites. This research laid the foundation and context for discussing management practices aimed at increasing food for bobwhites, such as supplemental feeding, from a scientific perspective. Researchers, managers, and students of bobwhites are much indebted to the work and research of these two scientists.

NUTRITION

It is generally known that breeding bobwhite hens have greater nutritional demands than non-breeding hens. Eggs are high in protein, and eggshells are high in calcium. Egg production requires energy above that necessary for basic sustenance, and bobwhites produce large clutches consisting of about 12 eggs. In addition, bobwhites can renest up to three or four times in a single season in an attempt to hatch at least one successful nest. It is not difficult then to understand why a breeding bobwhite would have much greater nutritional demands than non-breeding bobwhites. The important question, however, is how much greater?

Nestler conducted experiments in which he monitored date of laying, laying-season length, egg production (number per hen), and numerous other variables as a function of protein level ranging from 13% to 29% in increments of 2%. Egg production occurred at all protein levels. However, hens that were fed the 23% protein diet laid about twice as many eggs (69 eggs per hen) as hens that were fed the 13% protein diet (36 eggs per hen). Egg production did not increase much beyond this level of protein. Egg laying also began earlier and ended later (i.e., longer laying season) as the protein level increased up to 23%. Maximum egg production therefore appeared to occur on a diet consisting of 23% protein with minimal improvements in productivity with increasing protein levels.

Nestler followed his initial work on protein requirements with the requirements of two minerals (calcium and phosphorus) important for reproduction. These two minerals are important not only in skeletal formation but also in other bodily functions, such as regulation of ion concentrations in blood and tissues and egg formation. He evaluated egg production at varying ratios of calcium to phosphorus. His research indicated that maximum egg production and the least amount of defective eggshells occurred at 2.3% calcium and 1% phosphorus.

An interesting note in Nestler's research trajectory was his work on vitamin A. During the summer of 1940, a local breeder brought Nestler several bobwhites in poor condition, suffering from emaciation and severe ophthalmia (eye inflammation), both symptoms of vitamin A deficiency. This incident spurred a series of investigations on the requirements of vitamin A for bobwhites and its possible role in the boom-and-bust phenomenon of the species. Vitamin A is synthesized in the liver from its precursor carotene, a pigment in plants. Thus, vitamin A is found solely in the tissues of animals. Bobwhites, having a diet consisting mostly of plant materials, depend largely on plants for their source of carotene and ultimately vitamin A. Because less carotene is produced in drought-stressed plants, the hypothesis was that poor reproduction occurred during drought because of vitamin A deficiency.

Nestler determined that optimum egg production and egg hatchability occurred at a vitamin A level of 6,000 I.U. (International Unit—a unit of measurement in pharmacology for vitamins) per pound of feed. From the perspective of a daily requirement, bobwhites required at least 40 I.U. of vitamin A per day for survival during the winter and 100 I.U. per day for maximum body needs and optimum subsequent reproduction in the spring. Follow-up field investigations by various researchers in Alabama, Ohio, Texas, and Virginia provided no conclusive evidence

TABLE 2.1. Nutritional Requirements for Breeding, Non-breeding, and Growing Northern Bobwhites

| | Adults | | |
Nutrient	Breeding	Non-Breeding	Growing
Protein	23%	12%	28%
Calcium	2.3%		1.0%
Phosphorus	1.0%		0.75%
Energy (metabolizable energy/ounce food)	2.8 kcal/ounce		
Vitamin A (I.U. / lb. of feed)	6000 I.U.	2500 I.U.	3,000 I.U.

that vitamin A was deficient in wild bobwhites. However, researchers noted that finding evidence of vitamin A deficiency in the wild was difficult because bobwhites suffering from avitaminosis (any disease caused by long-term vitamin deficiency) would likely succumb to predation before being encountered afield. Vitamin A deficiency causes bobwhites to suffer from weakness, impaired eyesight, lack of alertness, and loss of speed, making them more vulnerable to predators. In an effort to address the possibility of vitamin A deficiency in wild bobwhites, Nestler investigated the carotene levels of several plants. His research indicated that yellow corn was an excellent source of the provitamin A and solely capable of sustaining bobwhites during winter with no average loss in body mass.

In summary, the work of Nestler indicated that reproducing bobwhites required about 23% protein, 2.3% calcium, 1% phosphorus, and 100 I.U. vitamin A per day for optimum reproduction. These requirements varied depending on the breeding status and age of the bobwhite (table 2.1).

BIOENERGETICS

Robert Robel and his students began their work on the energetics of bobwhites during the 1970s. Their research involved quantifying the energy requirements of male and female bobwhites under simulated seasonal conditions (10-hour or 15-hour photoperiod and temperatures ranging from 32°F to 104°F). The 10-hour photoperiod represented day length during the fall–winter period, whereas the 15-hour photoperiod represented day length during the spring–summer period. Long photoperiods (more than 12 hours) are required to stimulate bobwhite reproduction; thus, the 15-hour photoperiod was intended to stimulate reproduction and egg laying in female bobwhites.

Robel's findings indicated that gross energy intake decreased in both males and females with increasing temperature during both photoperiods. Body mass, however, tended to increase up to about 77°F, beyond which it decreased. Maximum body mass for both sexes occurred between 68°F and 77°F.

Regarding energetic requirements, both males and females had similar requirements at the 10-hour photoperiod (i.e., non-reproduction period). Males and non-laying females also had similar energetic requirements at the 15-hour photoperiod (i.e., reproduction period). However, as could be expected, laying females had higher energy requirements than either of these two groups. A laying female would require about 50 kcal per day at 77°F compared to about 35 kcal per day for a male or non-laying female. This difference in energy requirements (15 kcal per day) between a laying female and a male (or non-laying female) represented the energy required for egg production. The energetic cost of producing a bobwhite egg therefore averaged about 14–18 kcal per day. The highest egg production (one egg per 2.2 days) occurred at 59°F. Eggs were heaviest and had the greatest caloric value at 77°F. In summary, optimum reproductive performance of bobwhites was achieved at a temperature range approximately between 70°F and 77°F.

An interesting finding that emerged from this research was that males were able to withstand slightly higher temperatures (110°F–112°F) than females (106°F–108°F) before death occurred. This finding is significant because peak nesting season of bobwhites in Texas generally coincides with the hottest, driest months of the year (i.e., June–August). Given that the greatest body mass, highest quality eggs, and greatest egg production occur at relatively cool temperatures (70°F–77°F), high summer temperatures can have detrimental impacts on the reproduction of wild bobwhites. The general pattern of lower energy intake with increasing temperature, coupled with the greater sensitivity of females to heat stress, makes temperature an important factor to consider when discussing bobwhite nutrition, energetics, and reproduction on semiarid rangelands.

Nutritional Plane of Wild Bobwhites

Populations of bobwhites and other quail species in North America are well known for their drastic population fluctuations in response to variation in the amount of rainfall. Poor reproduction and low population numbers occur during drought; good reproduction and high numbers occur during years of abundant rainfall. An intuitive hypothesis that arises from these observations is that poor reproduction may occur during drought because of inadequate nutrition. This begs the question, are bobwhites capable of meeting their nutritional demands for reproduction in the wild?

Research indicates that the nutritional plane of wild bobwhites is adequate for reproduction during years of relatively low precipitation (as little as 16 inches), at least for crude protein and energy (but see below). Fred S. Guthery and his students documented that the diet of wild bobwhites in South Texas consisted of about 24% protein and provided an average of about 3.3 kcal/g dry weight of food consumed. Because breeding bobwhites require 23% protein and 2.8 kcal/g, the diet appeared adequate in crude protein and energy for reproduction. Calcium and phosphorous intake, however, was lower than recommended levels. Females selected a diet higher in calcium (1.34%) than males (0.79%), but calcium intake was still about 40% lower than the suggested 2.3%. Phosphorus intake was about 0.25% for both sexes and about 75% below the recommended 1%. Despite these apparent deficiencies, the bobwhite population under study still exhibited reasonable productivity (1.8 juveniles per adult). In addition, the region was known to support good densities of bobwhites (one or more bobwhites per acre). Several explanations were offered for the inconsistency between apparent nutrient deficiencies and reasonable productivity: 1) the recommended levels for optimum reproduction obtained using captive bobwhites did not relate well to wild bobwhites, 2) bobwhites mobilized internal stores to account for the difference, or 3) the recommendations represented optimum and not minimum levels necessary for reproduction. Indeed, egg laying is known to occur with as little as 1.6% calcium and 0.35% phosphorus. The work of Guthery indicated that crude protein and energy appeared adequate for reproduction and that marginal intake of calcium and phosphorus was not associated with reproductive failure. However, they noted that it was possible for the productivity and density of bobwhites to be higher if the nutritional plane was corrected for these two nutrients.

Even though the bobwhite diet may be adequate for reproduction during years of low precipitation relative to crude protein needs, it is still possible for certain amino acids, the building blocks of protein, to be limiting. There are 20 amino acids, of which 10 are considered essential for birds. Amino acids are considered essential when the body is unable to synthesize them or cannot synthesize them in adequate quantities and therefore must be obtained from the diet. Research from western Oklahoma supports Guthery's finding that the level of crude protein in the bobwhite diet appears adequate for reproduction. However, the research indicates the bobwhite diet can be deficient in one or more essential amino acids for growth and breeding. In the study, the sulfur-containing amino acids (methionine + cystine) were consistently the most limited. In addition, bobwhite diets dominated by seeds appeared more limited in essential amino acids than insect-dominated diets. These findings suggest that nutrient limitations may exist for essential amino acids despite adequate intake of crude protein.

One way bobwhites may deal with nutrient limitations is to adjust dietary intake to correct for such deficiencies. Research indicates that bobwhites, like most other avian species, are capable of detecting caloric deficiencies but not protein deficiencies. That is, bobwhites can increase food consumption to partially compensate for energy deficiencies but will not increase consumption in response to a low-protein diet. This dietary adjustment of increased consumption in response to a low-energy but not low-protein diet is a common behavioral response in both bobwhites and scaled quail. However, scaled quail reproduction and body mass are less affected by a low-protein and low-energy diet than bobwhites, suggesting that bobwhites have higher nutritional demands for reproduction. This lower sensitivity to nutritional deficiencies, coupled with their more diverse diet, may explain why scaled quail populations do not boom or bust as drastically as bobwhites.

In Summary

The general food habits of bobwhites and their nutritional requirements for optimum breeding are well documented. Research indicates that the nutritional plane of wild bobwhites is, generally speaking, adequate for reproduction. However, there are possible deficiencies or at least nutrient limitations that may exist, particularly for calcium, phosphorus, and sulfur-containing amino acids. These nutritional limitations do not appear to result in complete reproductive failure, but the magnitude of their effect on reproduction is unknown. It is possible that reproduction could be greater if these nutritional deficiencies were corrected.

CHAPTER THREE

Common Seeds Eaten by Bobwhites

Notes on Using This Section

Many seeds in this book are quite unique and relatively easy to identify. Some, on the other hand, may be more difficult to identify and could be confused with seeds of other species of plants in this book. Contained within the verbal description of each of the seeds are defining characteristics such as size, color, shape, texture, and other distinctive features that the seed may exhibit. There can be, however, wide variations among seeds originating from the same species of plant. Several factors that can influence these differences are weather, soil physical and chemical properties, and subspecies or varieties within a species. Different seed variations (if available) for each plant species were photographed to illustrate the variation within a species. To aid in identification, seeds removed from a crop may need to be lightly washed so that the seed coat is clean and easily seen. Also, a magnifying glass and ruler may be necessary for positive identification because several of the seeds photographed in this book are very small. For seeds that are very difficult to identify, another possible approach for identification, one that requires a little more time, is to plant the seed and identify the plant that grows from that seed. Protein content for each seed is also provided in this section, if known. In addition, distribution maps represent native and naturalized populations and may not reflect all areas where a species can be found. Below are some abbreviations that are used in this section:

spp.—plural for "species": denotes more than one species within a genus.
ssp.—subspecies: a taxonomic group that is a subdivision of a species, usually based on geographical distribution.
var.—variety: a taxonomic group consisting of members of a species that can interbreed but not such that all traits are always expressed.

COMMON SEEDS EATEN BY BOBWHITES

CYPERACEAE
Fringed Nutrush
Scleria ciliata Michx.

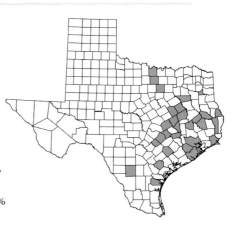

Occurred in about 3% of 33 crops examined from north-central Texas and comprised less than 1% of total seed volume. Outline oval with a protruding hilum. White to dark gray, covered with fine ridges, glossy. Length 2.7–3.1 mm; width 2–2.4 mm. Seeds contain about 10% protein.

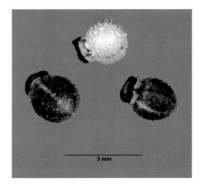

3 mm

Description of Plant

LIFEFORM: Native, perennial sedge
HEIGHT: To 3'
STEMS: Triangular, slender
BLADES: Linear, to 18" long, less than ½" wide
FLOWERS: Small, terminal
FRUITS: Achene, whitish
FLOWERING: April to October
HABITAT: Found on sandy soils in low areas, woods, and openings

Wild Oat
Avena fatua L.

Occurred in 2% of 200 crops examined from South Texas and comprised about 1% of total seed volume. Outline linear. Tan, papery, with one tip tapering sharply. Length 14–18 mm; width 2.5–3 mm. Seeds contain about 14.8% protein.

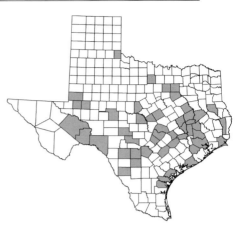

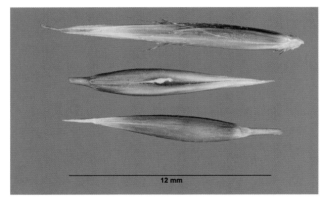

12 mm

Description of Plant

SYNONYM: Oats
LIFEFORM: Introduced, annual grass
HEIGHT: To 4′
STEMS: Erect, smooth
BLADES: Flat to twisting, hairless, to about ½″ wide
SEEDING: March to June
HABITAT: Wild oat is very similar to common oat (*A. sativa*) and is found along roadsides and in ditches.

Rescuegrass
Bromus catharticus Vahl

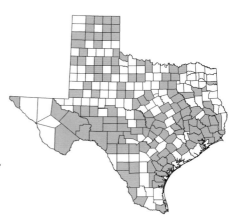

Occurred in about 64% of 33 crops examined from north-central Texas and comprised about 41% of total seed volume. Outline linear with one end tapering sharply. Tan, smooth with lengthwise ridges, dull. Length 12–15 mm; width 2–3 mm. Seeds contain about 18.2% protein.

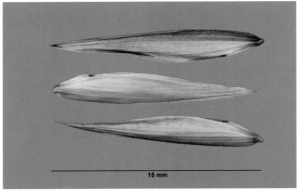

15 mm

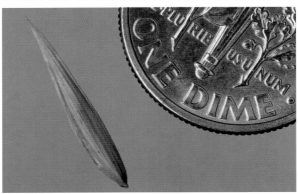

Description of Plant

SYNONYMS: Rescue brome, Prairie grass

LIFEFORM: Introduced, annual or perennial grass

HEIGHT: To 32"

STEMS: Erect, soft in new growth, in clumps

BLADES: Hairy or hairless, often with a yellowish band, to about ½" wide

SEEDING: February to May

HABITAT: Frequent on a variety of soils in ditches, fields, openings, and along roadsides

Coastal Sandbur

Cenchrus spinifex Cav.

Occurred in quail diets during the winter, summer, and fall for crops examined from South Texas. Outline teardrop-shaped, tapering to a fine point. Tan, smooth, dull. Length 4.3–4.9 mm; width 1.5–1.8 mm.

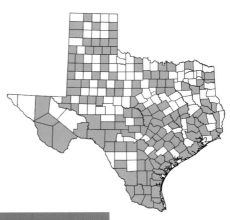

5 mm

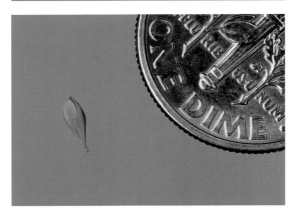

Description of Plant

SYNONYMS: Sandbur, Grassbur, Coast sandspur
LIFEFORM: Native, annual or perennial grass
HEIGHT: To 32"
STEMS: Erect or spreading
BLADES: Thin, flat, to about 7" long
SEEDING: Summer to fall
HABITAT: Common in pastures, ditches, along roads, and in disturbed
areas

Rosette Grass

Dichanthelium spp. (Hitchc. & Chase) Gould

Occurred in about 13% of 63 crops examined from South Texas and comprised about 19% of total seed volume. *Dichanthelium ovale* (pictured). Outline oval to almost circular. Light tan, smooth, glossy. Length 1.5 mm; width 1 mm.

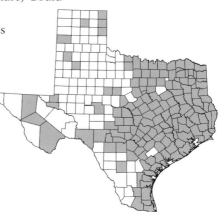

1 mm

Description of Plant

LIFEFORM: Native, perennial grass
HEIGHT: To about 20"
STEMS: Erect to spreading
BLADES: Thick, hairless, to just over ¼" wide
SEEDING: March to June and again in fall
HABITAT: Frequent on a variety of soils often in shaded woodland areas

Barnyardgrass
Echinochloa crus-galli (L.) P. Beauv.

Widely listed as important to bobwhites. Outline widely elliptic. Tan, sometimes with a long awn attached, lemma and palea are smooth and shiny, seed glossy. Length 3.5–4.2 mm; width 1.6–2 mm. Seeds contain about 12.1% protein.

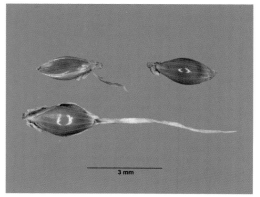

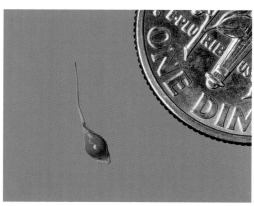

Description of Plant

SYNONYMS: Cockspur, Watergrass, Barnyard millet, Barn grass, Barn millet

LIFEFORM: Introduced, annual grass

HEIGHT: To 30"

STEMS: Erect, smooth, purplish, nodes slightly swollen

BLADES: Rough or sparsely hairy, to 16" long and ½" wide

SEEDING: July to November

HABITAT: Common on moist, fertile soils in ditches, roadsides, field edges, and disturbed areas

Witchgrass
Panicum capillare L.

Occurred in about 4% of 200 crops examined from South Texas and comprised about 1% of total seed contents. Outline elliptic. Tan to brown, smooth, glossy. Length 1.3 mm; width 0.8 mm.

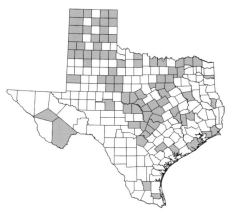

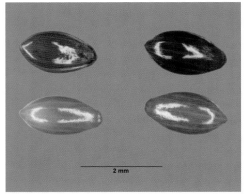

2 mm

Description of Plant

SYNONYMS: Common witchgrass, Old witchgrass, Tumblegrass
LIFEFORM: Native, annual grass
HEIGHT: To 32"
STEMS: Highly branched, spreading at base
BLADES: Flat, hairy, to 10" long and just over ½" wide
SEEDING: June to November
HABITAT: Frequent to infrequent on sandy and dry soils in fields, roadsides, and disturbed areas

Switchgrass
Panicum virgatum L.

Occurred in 23% of 100 crops examined from South Texas during fall and comprised about 1.2% of total seed contents. Outline oval. Tan to brown, usually with some dark gray or black markings, smooth, glossy. Length 1.8–2.5 mm; width 0.9–1.3 mm.

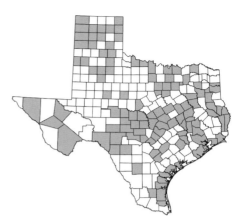

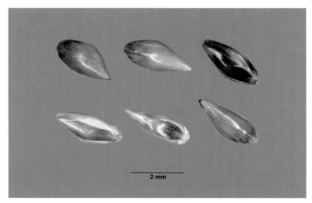

2 mm

Description of Plant

SYNONYMS: Tall prairiegrass, Wobsqua grass, Blackbent, Wild redtop
LIFEFORM: Native, perennial grass
HEIGHT: To 9'
STEMS: Stout, typically in large clumps
BLADES: Flat, firm, sometimes with light hairs, to 2' long
SEEDING: August to October
HABITAT: Frequent on moist soils, sometimes clay or sandy soils in
  openings

Brownseed Paspalum

Paspalum plicatulum Michx.

Comprised about 16% of total seed contents of 27 crops examined from South Texas during summer and fall. Outline oval. Dark brown, smooth, glossy. Lemma and palea light brown. Length 2.2–2.5 mm; width 1.2–1.4 mm.

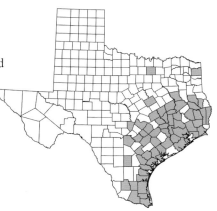

Description of Plant

LIFEFORM: Native, perennial grass
HEIGHT: To 40"
STEMS: Erect, clumped
BLADES: Firm, folded at base, often hairy above, to ¼" wide
SEEDING: June to November
HABITAT: Common on sandy and loamy soils in open oak woodlands and prairies

3 mm

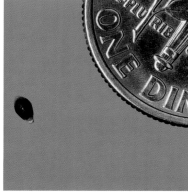

Thin Paspalum
Paspalum setaceum Michx.

Occurred in 65% of 51 crops examined from South Texas during fall and comprised about 5.3% of total seed contents. Outline oval. Tan, often with very small brown flecks. Length 1.5–1.8 mm; width 0.9–1.1 mm.

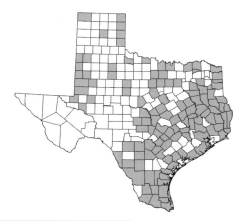

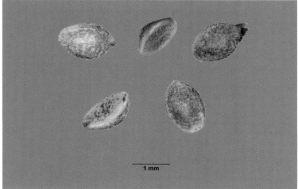

1 mm

Description of Plant

SYNONYM: Beadgrass

LIFEFORM: Native, perennial grass

HEIGHT: To 40"

STEMS: Clump-forming

BLADES: Varying in length, to ¾" wide, often hairy

SEEDING: May to October

HABITAT: Grows on a wide range of soils from sandy to clayey soils in ditches, disturbed areas, pastures, and woodland edges

PHOTO BY MARCIE O'CONNOR

Streambed Bristlegrass
Setaria leucopila (Scribn. & Merr.) K. Schum.

Occurred in 42% of 100 crops examined from South Texas during fall and
comprised about 4% of total seed
contents. Outline oblong. Pale
tan to light brown, one face
flat, the other with a prominent hump, smooth. Length
1.8–2 mm; width 0.9–1.1 mm.

2 mm

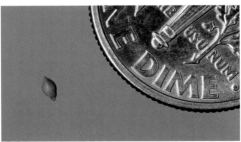

Description of Plant

SYNONYM: Plains bristlegrass

LIFEFORM: Native, perennial grass

HEIGHT: To 4'

STEMS: Erect, in dense clumps

BLADES: Flat or folded, thin, to 10" long

SEEDING: Late spring to fall

HABITAT: Found on sandy, loamy, and clayey soils in dry prairies and
   pastures, and beneath brush

Sorghum
Sorghum bicolor (L.) Moench

Occurred in about 51% of 51 crops examined from South Texas during fall and comprised about 45% of total seed volume. Outline circular to oval. Reddish-brown, flat area on one face, smooth, dull. Length 3.9–5.1 mm; width 3.1–4.2 mm. Seeds contain about 14% protein.

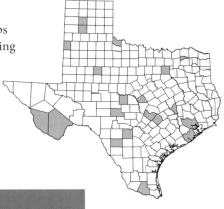

MAP DEPICTS ONLY WHERE SORGHUM IS NATURALIZED; IT IS WIDELY CULTIVATED.

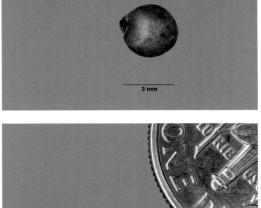

3 mm

Description of Plant

SYNONYMS: Grain sorghum, Milo, Maize
LIFEFORM: Introduced, annual grass
HEIGHT: To 10'
STEMS: Erect, stout
BLADES: Long, thin, to 2" wide
SEEDING: Summer to fall
HABITAT: Commonly planted in agricultural fields

Johnsongrass
Sorghum halepense (L.) Pers.

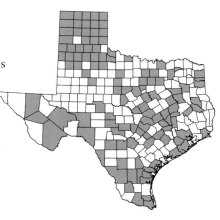

Occurred in about 14% of 91 crops examined from north-central Texas and comprised about 6% of total seed volume. Outline elliptic, often with one or two hairy, "horn-like" extensions along about half of the length of the seed. Mostly dark brown, smooth, glossy. Length 4.5–4.8 mm; width 1.5–2 mm. Seeds contain about 17.7% protein.

Description of Plant

LIFEFORM: Introduced, perennial grass
HEIGHT: To 6'
STEMS: Erect, woody
BLADES: Large, flat, usually hairless, to 18" long
SEEDING: May to November
HABITAT: Johnsongrass is often cultivated, grows on various soils, and is common along roadsides, ditches, and disturbed areas.

Common Wheat
Triticum aestivum L.

Occurred in about 4% of 167 crops examined from East Texas during winter and comprised about 1% of total seed contents. Outline oblong. Pale tan, smooth, with a deep groove running lengthwise along one face, and fine hairs on one tip. Length 6–7 mm; width 2.8–4 mm.

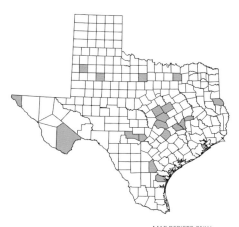

MAP DEPICTS ONLY WHERE WHEAT IS NATURALIZED; IT IS WIDELY CULTIVATED.

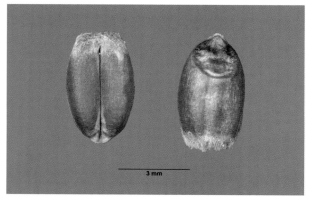

3 mm

Description of Plant

SYNONYMS: Wheat, Bread wheat
LIFEFORM: Introduced, annual grass
HEIGHT: To 40" tall
STEMS: Erect, stiff, hollow
BLADES: Thin, to about 1' long
SEEDING: March to May
HABITAT: Wheat is a cultivated cool-season crop and rarely grows in ditches and disturbed areas.

Fringed Signalgrass
Urochloa ciliatissima (Buckley) R. Webster

Occurred in about 54% of 108 crops examined from South Texas during summer and comprised about 2% of total seed contents. Outline oval. Whitish to tan, lemma and palea light green and hairy, seed smooth, dull. Length 3 mm; width 2.5–3 mm.

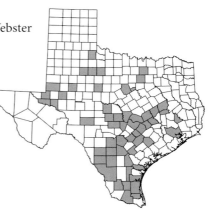

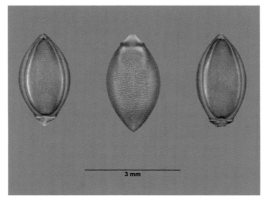

3 mm

Description of Plant

LIFEFORM: Native, perennial grass
HEIGHT: To 16"
STEMS: Erect to drooping, singly or in clumps
BLADES: Rough, to 8" long
SEEDING: April to June
HABITAT: Occurs on sandy soils in prairies and openings

Browntop Signalgrass
Urochloa fusca (Sw.) B. F. Hansen & Wunderlin

Occurred in 77% of 100 crops examined from South Texas during fall and comprised about 15% of total seed contents. Outline broadly elliptic. Tan, palea and lemma usually light green to tan and covering seed tightly, dull. Length 3 mm; width 2 mm.

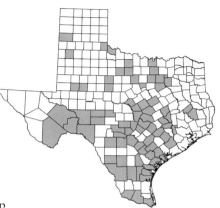

Description of Plant

SYNONYMS: Signalgrass, Browntop
LIFEFORM: Native, annual or perennial grass
HEIGHT: To 40"
STEMS: Erect, curving upward from base, hairy
BLADES: Flat, lightly hairy, to 12" long and ¾" wide
SEEDING: June to November
HABITAT: Frequent on moist soils in ditches, fields, and disturbed areas

3 mm

Texas Signalgrass
Urochloa texana (Buckley) R. Webster

Occurred in about 30% of 565
crops examined from southwest
Texas and comprised about 30%
of total contents. Outline elliptic
with one end tapering to a fine
point. Greenish to light tan,
with fine ridges, shiny. Length
2.9–4 mm; width 1.9–2 mm.

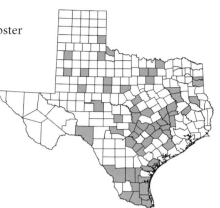

3 mm

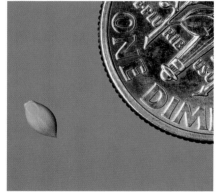

Description of Plant

SYNONYMS: Coloradograss, Texas panicum
LIFEFORM: Native, annual grass
HEIGHT: To 4'
STEMS: Stout, sometimes creeping, often clumping
BLADES: Firm, hairy, to 8" long
SEEDING: May to November
HABITAT: Found on moist soils in ditches, field borders, and disturbed
areas

Corn

Zea mays L.

Occurred in 23% of 44 crops examined from South Texas during winter and comprised about 18% of total seed contents. Outline oval to somewhat triangular. Yellow, smooth, shiny. Length 8–15 mm; width 7–9 mm.

MAP DEPICTS ONLY WHERE CORN IS NATURALIZED; IT IS WIDELY CULTIVATED.

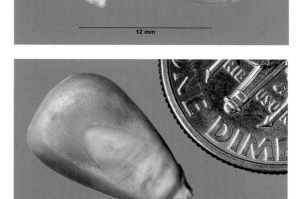

12 mm

RUSHES AND GRASSES

Description of Plant

SYNONYM: Maize
LIFEFORM: Introduced, annual grass
HEIGHT: To 6' or more
STEMS: Stout
BLADES: Prominently ribbed, broad
SEEDING: May to July
HABITAT: Corn is a cultivated crop, is widely planted, and rarely grows without cultivation.

ACANTHACEAE
Gregg's Tube Tongue
Justicia pilosella (Nees) Hilsenb.

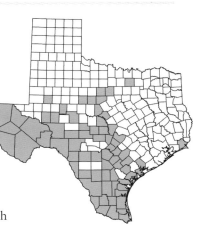

Occurred in about 18% of 11 crops examined from southwest Texas during fall and winter and comprised less than 1% of total seed and fruit volume. Outline circular with a small notch along seed margin. Orange to dark brown, flattened, covered with tiny bumps, dull. Length 2.8–3.5 mm; width 2.8–3.5 mm.

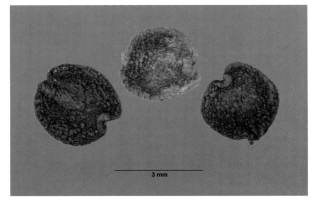

3 mm

Description of Plant

SYNONYMS: Hairy tubetongue, Tubetongue
LIFEFORM: Native, perennial forb
HEIGHT: To 1'
STEMS: Erect to drooping, hairy
LEAVES: Simple, opposite, oval, with prickles, to 1½" long
FLOWERS: Lavender, extending from a white floral tube
FRUITS: Capsule, four-seeded
FLOWERING: April to October
HABITAT: Found on various soils on hilltops, pastures, and along
streams.

AMARANTHACEAE
Carelessweed
Amaranthus palmeri S. Watson

Amaranthus spp. occurred in about 15% of 33 crops examined from north-central Texas and comprised about 1% of total seed volume. Outline oval to circular. Black to reddish-brown, smooth, glossy. Length 1–1.7 mm; width 0.8–1.5 mm. Seeds (*A. palmeri*) contain about 15.8% protein.

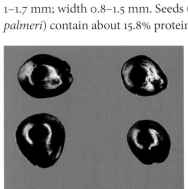

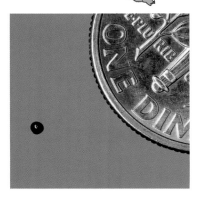

1 mm

Description of Plant

SYNONYMS: Pigweed, Palmer's amaranth
LIFEFORM: Native, annual forb
HEIGHT: Usually to about 40", sometimes to 6'
STEMS: Erect, smooth, often turning red
LEAVES: Simple, alternate, oval or lance-shaped, to 6" long
FLOWERS: Greenish-yellow
FRUITS: Utricle, small, one-seeded
FLOWERING: April to December
HABITAT: Common on silty, sandy, and gravelly soils in a variety of places including open areas, woods, ditches, stream banks, and disturbed areas

Cuman Ragweed
Ambrosia psilostachya DC.

Occurred in 100% of eight crops
examined from North Texas
and comprised 21% of total contents. Outline oval to triangular
with spines near the top. Brownish-
gray, mottled, surface rough with
wrinkles or fine hairs near the top.
Length 3.2–3.5 mm; width 1.7–2 mm.
Seeds contain about 12.1% protein.

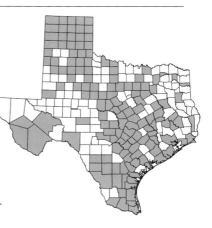

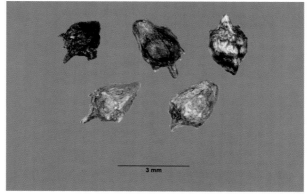

3 mm

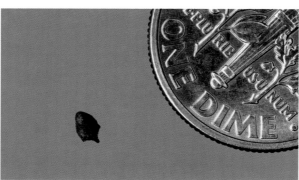

Description of Plant

SYNONYMS: Western ragweed, Perennial ragweed
LIFEFORM: Native, annual or perennial forb
HEIGHT: To 4'
STEMS: Erect, hairy
LEAVES: Simple, alternate, deeply lobed, to 2½" long
FLOWERS: White, small, drooping
FRUITS: Achene, one-seeded
FLOWERING: July to December
HABITAT: Common on a variety of soils in pastures, woods, stream bottoms, and disturbed areas

Great Ragweed
Ambrosia trifida L.

Occurred in about 4% of 56 crops examined from East Texas during winter and comprised less than 1% of total seed volume. Outline oval to triangular with spines near the top. Tan, papery, often with fine hairs near the top. Length 5.6–6.3 mm; width 3–3.8 mm. Seeds contain about 13.1% protein.

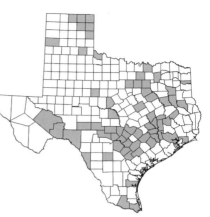

3 mm

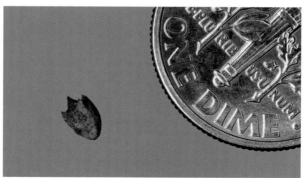

Description of Plant

SYNONYMS: Giant ragweed, Horsecane, Richweed, Blood ragweed, Buffaloweed

LIFEFORM: Native, annual forb

HEIGHT: To 9'

STEM: Angled, green, branching

LEAVES: Simple, opposite, usually three-lobed, 4–8" long

FLOWERS: Inconspicuous

FRUITS: Achene, woody, one-seeded

FLOWERING: August to December

HABITAT: Great ragweed is most common in disturbed areas, along stream bottoms, and on moist soils.

Common Sunflower
Helianthus annuus L.

Occurred in about 5% of 361 crops examined from east-central Texas and comprised about 2.3% of total contents. Outline somewhat triangular to oblong. Grayish brown or lighter, sometimes with light hairs, compressed. Length 4–4.9 mm; width 2.3–2.6 mm. Seeds contain about 13.8% protein.

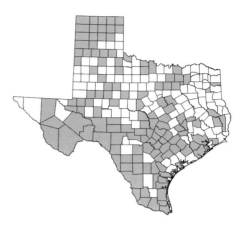

6 mm

Description of Plant

SYNONYMS: Sunflower, Annual sunflower, Wild sunflower
LIFEFORM: Native, annual forb
HEIGHT: To over 6'
STEMS: Erect, branching, with coarse hairs, sticky
LEAVES: Simple, alternate, triangular, rough, up to 1' long
FLOWERS: Bright yellow, with dark disk
FRUITS: Achene, lightly hairy
FLOWERING: March to December
HABITAT: Common sunflower occurs on soils ranging from clayey to sandy in a variety of habitats including openings, recently cultivated fields, roadsides, and streambeds.

Silverleaf Sunflower

Helianthus argophyllus Torr. & A. Gray

Occurred in about 5% of 63 crops examined from South Texas during winter and comprised about 4.2% of total seed contents. Outline triangular. Gray with black markings, fine longitudinal ridges, somewhat shiny. Length 3.2–4.2 mm; width 2–2.3 mm.

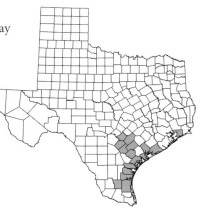

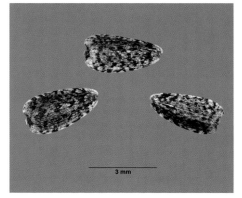

3 mm

Description of Plant

LIFEFORM: Native, annual forb
HEIGHT: To 12'
STEMS: Erect, silvery, rough, branching on upper half
LEAVES: Simple, alternate, silvery, wooly, to 12" long
FLOWERS: Yellow, with dark disk
FRUITS: Achene
FLOWERING: June to December
HABITAT: Frequent on deep sandy soils along the coast and in stream bottoms

Camphorweed

Heterotheca subaxillaris (Lam.) Britton & Rusby

Occurred in about 18% of 63 crops examined from South Texas during winter and comprised about 7.4% of total seed contents. Outline oval to oblong. Tan to light brown, margins slightly higher than center, smooth, dull. Length 1.7–2 mm; width 0.9–1.1 mm. Seeds contain about 13.9% protein.

Description of Plant

SYNONYM: Golden aster
LIFEFORM: Native, annual forb
HEIGHT: To 3'
STEMS: Rough, green to purple, hairy
LEAVES: Simple, alternate, rough, to 3½" long
FLOWERS: Yellow
FRUITS: Nutlet, one-seeded
FLOWERING: October to November
HABITAT: Common on sandy soils in openings, prairies, and disturbed areas

Golden Crownbeard
Verbesina encelioides (Cav.) Benth. & Hook. f. ex A. Gray

Occurred in 30% of 51 crops examined from South Texas during winter and comprised about 1.7% of total seed contents. Outline triangular, often with two stiff "hairs" extending from widest end. Edges tan, center black, covered with fine hairs. Length 5.7–7.2 mm; width 2.5–4.7 mm. Seeds contain about 28.8% protein.

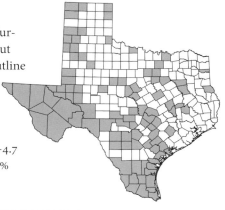

Description of Plant

SYNONYM: Cowpen daisy
LIFEFORM: Native, annual forb
HEIGHT: To 3'
STEMS: Erect, green
LEAVES: Simple, alternate, hairy, 3–4" long
FLOWERS: Yellow, with a yellow center
FRUITS: Achene, flattened
FLOWERING: February to December
HABITAT: Common on sandy soils in fields, prairies, and disturbed areas

BRASSICACEAE
Western Tansymustard
Descurainia pinnata (Walter) Britton

Occurred in about 43% of 63 crops
examined from South Texas
during spring and comprised
about 7% of total seed contents.
Outline oblong. Light brown, with
small pits covering the surface and
grooves running lengthwise along
the seed. Length 0.9–1.1 mm; width
0.5–0.6 mm.

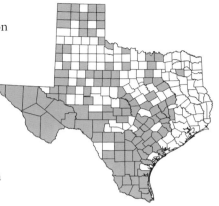

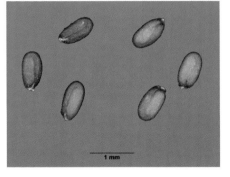

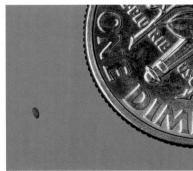

Description of Plant

SYNONYM: Pinnate tansymustard
LIFEFORM: Native, annual, biennial, or perennial forb
HEIGHT: To 30"
STEMS: Erect, typically single-stemmed
LEAVES: Compound, alternate, at base of stem
FLOWERS: Pale yellow
FRUITS: Capsule
FLOWERING: January to April
HABITAT: Western tansymustard grows well on sandy soils in prairies,
    disturbed areas, and openings.

COMMELINACEAE
Whitemouth Dayflower
Commelina erecta L.

Occurred in 67% of 108 crops examined from South Texas during
summer and comprised about
9.4% of total seed contents. Outline circular to oval. Mostly black
to dark brown, somewhat smooth,
dull, with a curved hilum on the flat-
ter face. Length 2.2–3 mm; width 2.1–2.3
mm. Seeds contain about 15.4% protein.

Description of Plant

SYNONYMS: Erect dayflower, Widow's tears
LIFEFORM: Native, perennial forb
HEIGHT: To over 2'
STEMS: Erect or spreading, smooth
LEAVES: Simple, alternate, to 6" long and 1" wide
FLOWERS: Violet to blue, less than 1" long, wilting at midday
FRUITS: Capsule
FLOWERING: March to December
HABITAT: Found on a variety of soil textures in prairies, openings,
    stream bottoms, and along roads

PHOTO BY TIMOTHY E. FULBRIGHT

Texas Bindweed
Convolvulus equitans Benth.

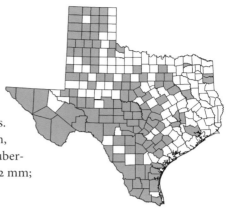

Occurred in 50% of 100 crops examined from South Texas during winter and comprised about 63% of total seed contents. Outline oval. Tan to dark brown, often with a small whitish protuberance, smooth, dull. Length 1.8–2 mm; width 1–1.3 mm.

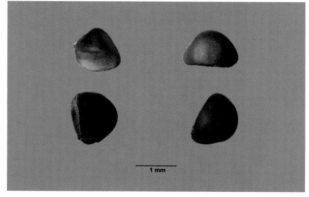

1 mm

Description of Plant

SYNONYM: Gray bindweed
LIFEFORM: Native, annual or perennial vine
HEIGHT: To about 6″, mostly trailing on ground
STEMS: Slender, trailing or twining, very hairy, to 6′ long
LEAVES: Simple, alternate, hairy, deeply lobed, to 3″ long
FLOWERS: White, often with a purple center, funnel-shaped, to 1½″
  wide
FRUITS: Capsule
FLOWERING: April to October
HABITAT: Frequent on sandy or rocky soils in prairies, openings, and
  disturbed areas

EUPHORBIACEAE
Cardinal's Feather
Acalypha radians Torr.

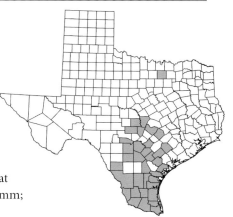

Occurred in 11% of 91 crops
examined from southwest
Texas during fall and winter
and comprised less than 1%
of total seed and fruit contents.
Outline oval. Gray, with a thin,
light-colored caruncle, somewhat
shiny. Length (with caruncle) 2 mm;
width 1 mm.

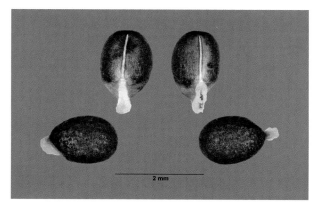

2 mm

Description of Plant

SYNONYM: Round copperleaf
LIFEFORM: Native, perennial forb
HEIGHT: To 16"
STEMS: Low-spreading, hairy
LEAVES: Simple, alternate, hairy, deeply lobed, to about 4¾" long
FLOWERS: Red
FRUITS: Capsule, three-seeded
FLOWERING: March to December
HABITAT: Common on sandy or gravelly soils in openings, prairies, and on island dunes

MALE PLANT OF ACALYPHA RADIANS

Low Silverbush

Argythamnia humilis (Engelm. & A. Gray) Müll. Arg.

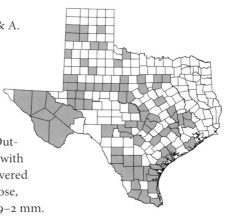

Occurred in about 40% of 91 crops examined from southwest Texas during fall and winter and comprised about 2% of total seed and fruit volume. Outline circular. Dark brown, often with symmetrical black markings, covered with very fine hairs, nearly globose, dull. Length 2–2.3 mm; width 1.9–2 mm.

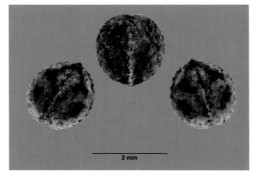

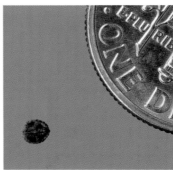

Description of Plant

SYNONYM: Wild mercury
LIFEFORM: Native, perennial forb or subshrub
HEIGHT: To 2'
STEMS: Woody base, erect to reclining, often hairy
LEAVES: Simple, alternate, hairy, to about 2" long
FLOWERS: Yellow-green, small
FRUITS: Capsule, three-seeded
FLOWERING: February to December
HABITAT: Common on a variety of soils in openings, prairies, and disturbed areas

Texas Bullnettle
Cnidoscolus texanus (Müll. Arg.) Small

Occurred in 100% of eight crops examined from North Texas and comprised about 66% of total contents. Outline oblong to oval. Gray to tan with light brown mottling, hilum raised, smooth, shiny. Length 14–15 mm; width 8–9 mm.

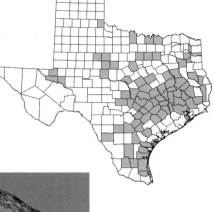

12 mm

Description of Plant

SYNONYMS: Mala mujer, Treadsoftly
LIFEFORM: Native, perennial forb
HEIGHT: To over 30"
STEMS: Thick, covered with stinging bristles
LEAVES: Simple, alternate, deeply lobed, with bristles, to 6" long
FLOWERS: White
FRUITS: Capsule, three-seeded
FLOWERING: March to November
HABITAT: Found on deep sandy soils in prairies and openings

Hogwort
Croton capitatus Michx.

Occurred in about 88% of 69 crops examined from South Texas during fall and comprised about 50% of total seed contents. Outline circular to oval, often with a caruncle at the tip. Tan with dark brown spots, glossy, caruncle light-colored. Length 4.7–5 mm; width 4.6–4.8 mm. Seeds contain about 20.9% protein.

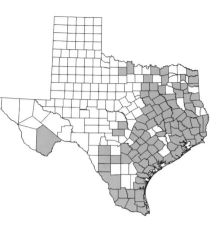

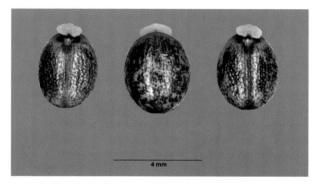

Description of Plant

SYNONYMS: Wooly croton, Doveweed, Goatweed
LIFEFORM: Native, annual forb
HEIGHT: To 4'
STEMS: Erect, branching, grayish-green, hairy
LEAVES: Simple, alternate, wooly, silvery, to about 5" long
FLOWERS: White
FRUITS: Capsule, three-lobed, each with one seed
FLOWERING: May to December
HABITAT: Common on sandy soils in prairies, openings, and disturbed
    areas

Vente Conmigo

Croton glandulosus L.

Occurred in about 52% of 69 crops
examined from South Texas
during summer and comprised
about 26% of total seed contents.
Outline oval with a pointed tip.
Tan with brown mottling, shiny,
often with a light-colored caruncle.
Length 3.3–3.7 mm; width 2.1–2.6 mm.

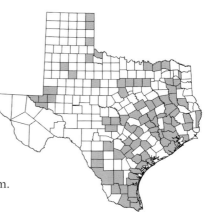

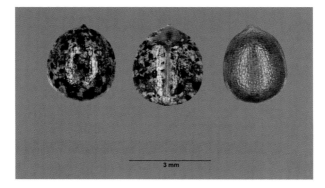

3 mm

Description of Plant

SYNONYMS: Tropic croton, Sand croton, Northern croton
LIFEFORM: Native, annual forb
HEIGHT: To 30"
STEMS: Erect with star-shaped hairs
LEAVES: Simple, alternate, toothed margins, to 3" long
FLOWERS: White, in small clusters
FRUITS: Capsule, three-seeded
FLOWERING: March to November
HABITAT: Varieties of this species are most common on sandy soils, but the plant is also found on loamy, clayey, or caliche soils in openings and prairies.

Prairie Tea
Croton monanthogynus Michx.

Occurred in about 2% of 200 crops examined from South Texas and comprised about 1% of total seed contents. Outline oval with a small pointed tip. Dark brownish-gray, smooth, shiny, caruncle light-colored. Length 2.5–3.2 mm; width 2–2.5 mm.

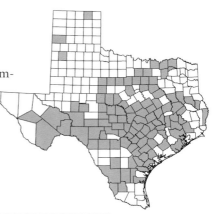

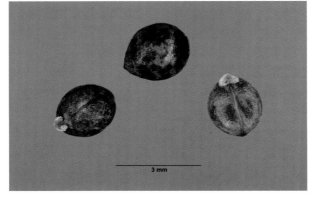

3 mm

Description of Plant

SYNONYMS: One-seeded croton, Lemon spurge
LIFEFORM: Native, annual forb
HEIGHT: To 16"
STEMS: Erect, branching, with scattered hairs
LEAVES: Simple, alternate, elliptic, to 2" long
FLOWERS: White, small, in clusters
FRUITS: Capsule, one-seeded
FLOWERING: April to December
HABITAT: Common on heavy loams, caliche, and clays in openings, prairies, and disturbed areas

Gulf Croton
Croton punctatus Jacq.

Occurred in about 11% of 565 crops examined from southwest Texas and comprised about 5.6% of total contents. Outline circular to oval. Light brown to black with various markings, smooth, some-what shiny. Length 4.2–5 mm; width 3.3–3.8 mm.

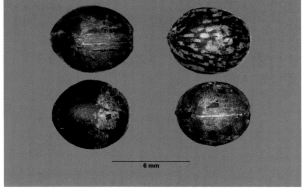

6 mm

Description of Plant

SYNONYM: Beach tea

LIFEFORM: Native, perennial forb or subshrub

HEIGHT: To 18"

STEMS: Erect, widely branching

LEAVES: Simple, alternate, grayish with brown spots underneath,
to 2" long

FLOWERS: Whitish, small

FRUITS: Capsule, three-seeded

FLOWERING: March to December

HABITAT: Common on deep, sandy soils all along the coast

Texas Croton

Croton texensis (Klotzsch) Müll. Arg.

Occurred in about 4% of 565 crops examined from southwest Texas and comprised 5.6% of total contents. Circular to broadly ovate with a short, blunt tip. Brown to gray, solid or mottled, smooth, shiny. Length 4–6 mm; width 3.5–5 mm. Seeds contain about 25% protein.

Description of Plant

SYNONYMS: Tinajera, Doveweed

LIFEFORM: Native, annual forb

HEIGHT: To 32"

STEMS: Erect, branching above, hairy

LEAVES: Simple, alternate, to 4" long and about 1" wide

FLOWERS: Small, whitish

FRUITS: Capsule, three-lobed, each with one seed

FLOWERING: May to November

HABITAT: Common on sandy soils in openings, prairies, and stream
   bottoms

Snow on the Prairie
Euphorbia bicolor Engel. & A. Gray

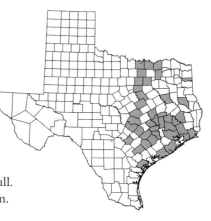

Occurred in about 42% of 91 crops examined from north-central Texas and comprised about 22% of total seed volume. Outline circular to broadly ovate. Greenish-gray with a fairly symmetrical pattern of ridges and the tip tapering to a broad, flat point, dull. Length 4.2–4.5 mm; width 3.4–4 mm. Seeds contain about 23.1% protein.

4 mm

Description of Plant

LIFEFORM: Native, annual forb

HEIGHT: To 3'

STEMS: Erect, wooly

LEAVES: Simple, alternate, oval, white margins on upper leaves, 1–3" long

FLOWERS: White

FRUITS: Capsule, three-lobed, each lobe with one seed, rounded, wooly

FLOWERING: July to October

HABITAT: Occurs on a wide range of soils and can be found in openings and prairies

Smartweed Leaf-flower
Phyllanthus polygonoides Nutt. ex Spreng.

Occurred in about 9% of 91 crops examined from southwest Texas during fall and winter and comprised less than 1% of total seed and fruit contents. Outline crescent-shaped to oval. Dark brown to black, curved on one face, flat on the other two faces, margins prominent, dull. Under high magnification, very small bumps can be seen. Length 1–1.5 mm; width 1 mm.

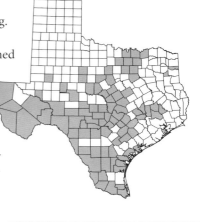

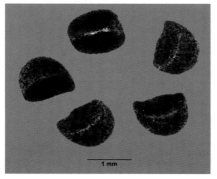

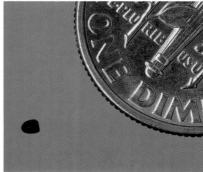

Description of Plant

SYNONYM: Knotweed leaflower
LIFEFORM: Native, perennial forb or subshrub
HEIGHT: To 16", usually less than 4"
STEMS: Grayish-green, smooth
LEAVES: Simple, alternate, about ½" long
FLOWERS: Greenish-yellow
FRUITS: Capsule
FLOWERING: March to November
HABITAT: Common on sandy loam or caliche in openings, woods, and disturbed areas

FABACEAE
Smallflowered Milkvetch
Astragalus nuttallianus DC.

Occurred in about 34% of 91 crops
examined from southwest Texas during fall and winter and
comprised about 2% of total seed
and fruit volume. Outline square
to rectangular, rarely triangular, with
a small notch on one margin. Tan to
reddish-brown, spotted, flattened, mostly
dull. Length 2–3 mm; width 2 mm. Seeds
contain about 35% protein.

1 mm

Description of Plant

SYNONYMS: Nuttal milkvetch, Turkeypea
LIFEFORM: Native, annual or perennial forb
HEIGHT: To 12″
STEMS: Erect to spreading
LEAVES: Compound, alternate, 7–19 leaflets, leaflets to ⅝″ long
FLOWERS: White, pink, lavender, or blue
FRUITS: Legume, slender, curving
FLOWERING: February to May
HABITAT: Frequent on soils ranging from sandy to clayey in prairies, openings, and disturbed areas

Spurred Butterfly Pea
Centrosema virginianum (L.) Benth.

Occurred in 24% of 167 crops examined from East Texas during winter and comprised about 4.4% of total contents. Outline oblong. Gray to reddish-brown with black mottling, hilum pale, surrounded by a dark band, smooth, glossy. Length 4–5 mm; width 2.2–2.8 mm.

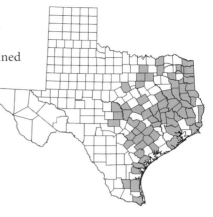

3 mm

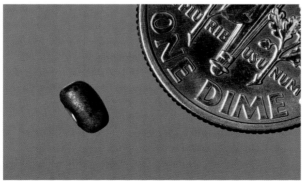

Description of Plant

SYNONYM: Butterfly pea

LIFEFORM: Native, perennial vine or forb

HEIGHT: Low, usually twining, sometimes trailing

STEMS: Twining, slightly hairy, to 6' long

LEAVES: Compound, alternate, three leaflets, leaflets linear or elliptic, to 2½" long

FLOWERS: Lavender to purple

FRUITS: Legume, flattened, with 4–10 seeds

FLOWERING: July to August

HABITAT: Found on sandy soils in forest openings, oak woods, and along the coast

Woodland Sensitive Pea

Chamaecrista calycioides (DC. ex Collad.) Greene

Occurred in about 11% of 63 crops examined from South Texas during winter
and comprised about 4.2% of total seed
contents. Outline square or rectangular, sometimes triangular.
Grayish to light brown, flattened,
with a pointed tip at one corner,
mostly shiny. Length 2.2–2.8 mm;
width 1.5–2.2 mm.

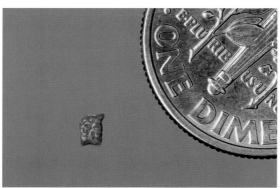

Description of Plant

LIFEFORM: Native, perennial forb
HEIGHT: To about 1'
STEMS: Low, hairy, to 20" long
LEAVES: Compound, alternate, leaflets to ½" long
FLOWERS: Yellow, with petals to ½" long
FRUITS: Legume, hairy
FLOWERING: July to August
HABITAT: Frequent on sands in openings and prairies

Partridge Pea
Chamaecrista fasciculata (Michx.)
Greene

Occurred in about 4% of 200 crops
examined from South Texas
and comprised about 7.5%
of total seed contents. Outline
diamond-shaped to nearly rectangular. Dark brown to black, with rows
of circular pits, flattened, shiny. Length
4–5 mm; width 2–3.3 mm. Seeds contain about 27% protein.

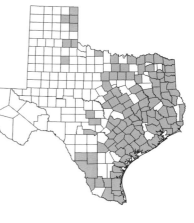

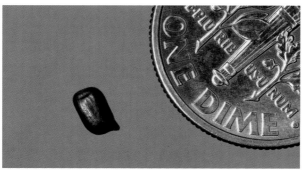

Description of Plant

SYNONYMS: Prairie senna, Showy partridge pea, Sleeping plant
LIFEFORM: Native, annual forb
HEIGHT: To 3'
STEMS: Erect or drooping, branching, hairy
LEAVES: Compound, alternate, 16–30 leaflets, leaflets to about ½" long
FLOWERS: Yellow
FRUITS: Legume, flattened
FLOWERING: June to December
HABITAT: Common on sandy soils in openings, prairies, dunes, and disturbed areas

FABACEAE

Texas Sensitive Pea
Chamaecrista flexuosa (L.) Greene

Occurred in quail diets during the summer for crops examined from South Texas. Outline some-what diamond-shaped, with one end pointed, flattened. Tan with light brown spots, smooth, dull. Length 4.2–4.6 mm; width 3–3.5 mm.

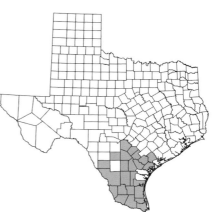

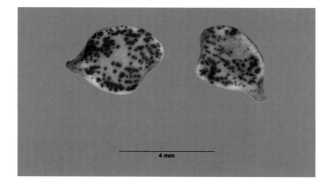

4 mm

Description of Plant

SYNONYM: Texas senna
LIFEFORM: Native, perennial forb
HEIGHT: To about 1½'
STEMS: Trailing, hairy, to 2' long
LEAVES: Compound, alternate, 20–32 leaflets
FLOWERS: Yellow
FRUITS: Legume, flattened
FLOWERING: April to September
HABITAT: Frequent on sandy or caliche soils in openings and
  disturbed areas

Sensitive Partridge Pea
Chamaecrista nictitans (L.) Moench

Occurred in 23% of 100 crops examined from South Texas during fall and comprised about 1.3% of total seed contents. Outline rectangular. Brown to black, with rows of circular pits, flattened, somewhat shiny. Length 2.5–3 mm; width 1–2.5 mm.

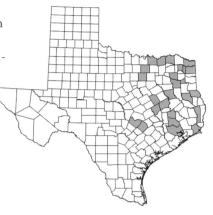

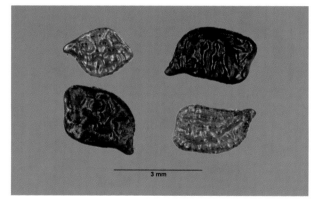

3 mm

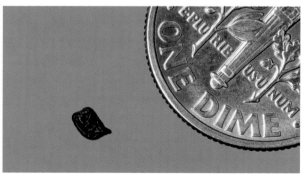

Description of Plant

SYNONYMS: Sensitive pea, Small partridge pea, Wild sensitive plant

LIFEFORM: Native, annual or perennial forb

HEIGHT: To 16"

STEMS: Slender, branching, slightly hairy

LEAVES: Compound, alternate, 20–40 leaflets, leaflets to about ¾" long

FLOWERS: Yellow

FRUITS: Legume, erect, flattened, usually with 5–10 seeds

FLOWERING: September to October

HABITAT: Frequent on sandy soils in openings, woodlands, and disturbed areas

Wedgeleaf Prairie Clover
Dalea emarginata (Torr. & A. Gray) Shinners

Occurred in quail diets during the winter, spring, summer, and fall for crops examined from South Texas. Outline oval to kidney-shaped. Light green to brown, smooth, somewhat shiny. Length 2 mm; width 1 mm. Seeds contain about 36.9% protein.

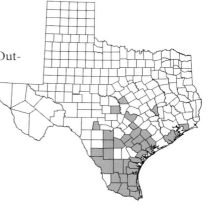

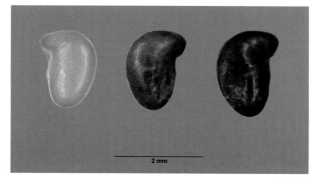

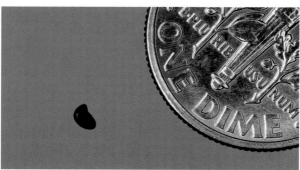

Description of Plant

SYNONYMS: Prairie clover, Dalea
LIFEFORM: Native, annual forb
HEIGHT: To 20″
STEMS: Erect, smooth
LEAVES: Compound, alternate, 13–17 leaflets, smooth
FLOWERS: Rose to purple
FRUITS: Legume, hairy
FLOWERING: March to December
HABITAT: Common on sandy, loamy, and clayey soils in openings and
disturbed areas

Illinois Bundleflower

Desmanthus illinoensis (Michx.) MacMill. ex B. L. Rob. & Fernald

Often listed as an important bobwhite food. Outline triangular, elliptic, or diamond-shaped. Dark orange to brown, with a dark crescent-shaped line on each face, smooth. Length 3.3–3.8 mm; width 2.3–2.6 mm. Seeds contain about 31% protein.

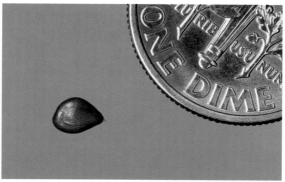

Description of Plant

SYNONYMS: Prairie mimosa, Prairie bundleflower, Prickleweed
LIFEFORM: Native, perennial forb or subshrub
HEIGHT: To 40"
STEMS: Erect or spreading
LEAVES: Compound, alternate, leaves to 3½" long
FLOWERS: White
FRUITS: Legume, flattened, curved
FLOWERING: May to June
HABITAT: Infrequent on clayey or caliche soils in openings and
　　ditches, or near water

PHOTO BY ANDY AND SALLY WASOWSKI. COURTESY OF LADY BIRD JOHNSON WILDFLOWER CENTER

Wild Tantan
Desmanthus virgatus (L.) Willd.

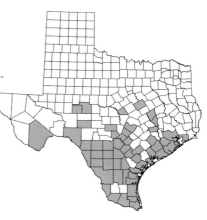

Occurred in about 35% of 91 crops examined from southwest Texas during fall and winter and comprised about 9% of total seed and fruit volume. Outline oval to almost circular. Orange to reddish-brown, flattened at margins, darker crescent-shaped line on each face, smooth, shiny. Length 2.6–3 mm; width 2.4–2.8 mm. Seeds contain about 33.5% protein.

Description of Plant

SYNONYMS: Bundleflower, Creeping bundleflower, Prostrate bundleflower

LIFEFORM: Native, perennial forb or subshrub

HEIGHT: To about 2'

STEMS: Woody, spreading or creeping, angled, smooth

LEAVES: Compound, alternate, leaflets to about ¼" long

FLOWERS: Greenish-white

FRUITS: Legume

FLOWERING: April to December

HABITAT: Common on a variety of soils in prairies, openings, and disturbed areas

Stiff Ticktrefoil

Desmodium obtusum (Muhl. ex Willd.) DC.

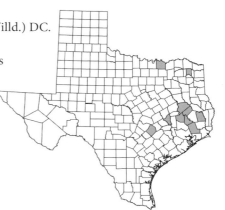

Occurred in about 44% of 36 crops examined from north-central Texas during winter and comprised about 48% of total contents. Outline oval. Brown, compressed, hilum small and circular, smooth, somewhat shiny. Length 2.5–3 mm; width 1.8–2 mm.

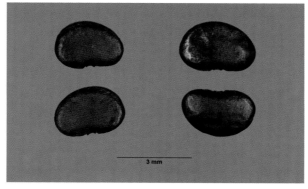

3 mm

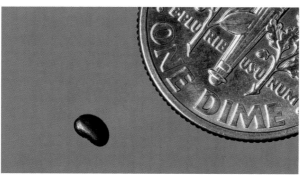

Description of Plant

SYNONYMS: Stiff tickclover, Panicled tick trefoil, Beggar lice, Beggar tick, Sticktight

LIFEFORM: Native, perennial forb

HEIGHT: To 5'

STEMS: Erect, branching above, leaning at top

LEAVES: Compound, alternate, leaflets hairy underneath, to about 2¼" long

FLOWERS: Pink to purple

FRUITS: Legume, flattened, 1–3 rounded segments, with sticky hairs

FLOWERING: June to September

HABITAT: Found on dry sandy or rocky soils in forest openings, fields, and roadsides

Hoary Milkpea

Galactia canescens Benth.

Occurred in about 26% of 69 crops examined from South Texas during fall and comprised about 6.1% of total seed contents. Outline kidney-shaped to oval. Cream-colored, brown, or reddish with mottling, sometimes black, hilum prominent, surrounded by a dark ring, smooth, shiny. Length 5.5–8 mm; width 4–6 mm.

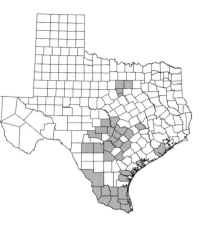

6 mm

Description of Plant

LIFEFORM: Native, perennial forb
HEIGHT: Creeping along ground
STEMS: Trailing
LEAVES: Compound, alternate, three leaflets, leaflets to just over
1" long
FLOWERS: Rose to purple
FRUITS: Legume, peanut-like, hairy, usually one-seeded
FLOWERING: April to October
HABITAT: Frequent on sandy soils in prairies and openings

Gray's Milkpea
Galactia heterophylla A. Gray

Occurred in about 17% of 69 crops examined from South Texas during summer and comprised about 4.3% of total seed volume. Outline kidney-shaped to almost rectangular. Red to gray with black mottling, smooth, shiny. Length 4.8–7 mm; width 3.2–5 mm.

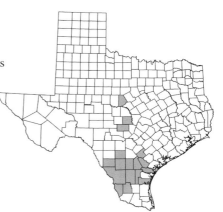

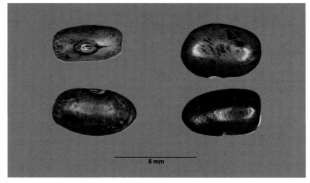

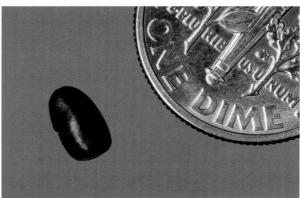

Description of Plant

SYNONYM: Varileaf milkpea
LIFEFORM: Native, perennial forb
HEIGHT: To 16"
STEMS: Trailing, to 2' long
LEAVES: Compound, alternate, with 3–5 leaflets, silvery, leaflets to about 1½" long
FLOWERS: Rose to lavender
FRUITS: Legume
FLOWERING: April to September
HABITAT: Frequent on sandy, loamy, or caliche soils in pastures and roadsides

PHOTO BY BILL CARR, COURTESY OF LADY BIRD JOHNSON WILDFLOWER CENTER

Eastern Milkpea

Galactia regularis (L.) Britton, Sterns & Poggenb.

Occurred in about 27% of 56 crops examined from East Texas during winter and comprised about 1% of total seed volume. Outline oval to kidney-shaped. Reddish-brown with black mottling, hilum surrounded by darker ring, smooth, shiny. Length 3.2–4.7 mm; width 2.5–3.1 mm.

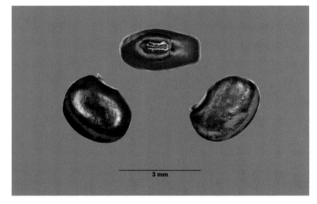

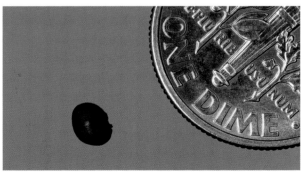

Description of Plant

SYNONYM: Milkpea
LIFEFORM: Native, perennial forb
HEIGHT: Creeping along ground
STEMS: Spreading
LEAVES: Compound, alternate, three leaflets, leaflets to about 2" long
FLOWERS: Violet to purple
FRUITS: Legume with stiff hairs
FLOWERING: June to August
HABITAT: Infrequent on dry sandy soils

Coastal Indigo
Indigofera miniata Ortega

Occurred in quail diets during the winter, spring, summer, and fall for crops examined from South Texas. Outline rectangular to oblong. Greenish, orange, or dark red, typically with black mottling, smooth, dull. Length 2–2.5 mm; width 1.3–1.5 mm. Seeds contain about 39.9% protein.

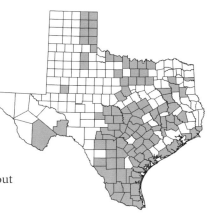

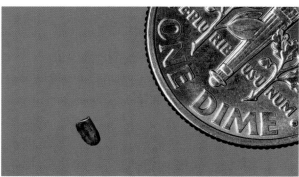

Description of Plant

SYNONYMS: Western indigo, Scarlet pea
LIFEFORM: Native, perennial forb
HEIGHT: To about 6"
STEMS: Trailing, to 3' long
LEAVES: Compound, alternate, hairy, grayish, 5–9 leaflets, leaflets to about 1" long
FLOWERS: Pink to red
FRUITS: Legume, hairy
FLOWERING: April to October
HABITAT: Common on sandy loams and sandy soils in prairies, openings, and dunes

Korean Clover
Kummerowia stipulacea (Maxim.) Makino

Occurred in 100% of three crops examined from North Texas and comprised about 37% of total contents. Outline oval with a small notch at one end. Dark red to black to pale, smooth, shiny. Length 1.7–2.2 mm; width 1.4–1.7 mm.

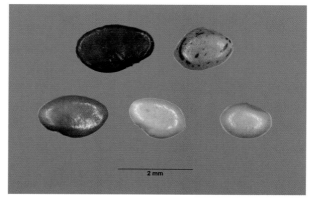

2 mm

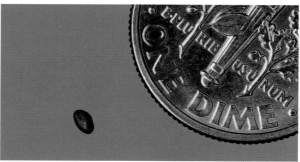

Description of Plant

SYNONYM: Korean lespedeza
LIFEFORM: Introduced, annual forb
HEIGHT: To 16", often forming mats
STEMS: Erect or reclining
LEAVES: Compound, alternate, three leaflets, leaflets to about ½" long
FLOWERS: Pink
FRUITS: Legume, with rounded tip
FLOWERING: June to September
HABITAT: Korean clover, often cultivated, is found on sandy soils and along roadsides

Slender lespedeza
Lespedeza virginica (L.) Britton

Occurred in about 18% of 56 crops examined from East Texas during winter and comprised about 2% of total seed volume. Outline oval to kidney-shaped. Reddish-brown, smooth, somewhat shiny, hilum small and circular. Length 2.5–3 mm; width 1.7–2 mm.

Description of Plant

SYNONYMS: Lespedeza, Slender bush clover, Bush clover
LIFEFORM: Native, perennial forb
HEIGHT: To 32″
STEMS: Often clustered, branched
LEAVES: Compound, alternate, three leaflets, hairy underneath
FLOWERS: Pink to purple with white
FRUITS: Legume, flattened
FLOWERING: August to October
HABITAT: Frequent on dry sandy soils in open woodlands and
roadsides

PHOTO BY DAN TENAGLIA. COPYRIGHT BY KAREN TENAGLIA

FABACEAE

Texas Lupine
Lupinus texensis Hook.

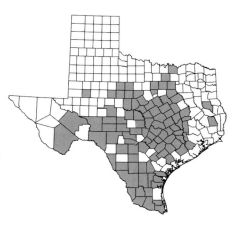

Occurred in 8% of 200 crops examined from South Texas and comprised less than 1% of total seed contents. Outline diamond-shaped to almost circular. Tan to gray with mottling, hilum sunken in protruding tip, smooth, dull. Length 5–5.8 mm; width 4.1–5 mm. Seeds contain about 34.4% protein.

4 mm

Description of Plant

SYNONYMS: Texas bluebonnet, Bluebonnet
LIFEFORM: Native, annual forb
HEIGHT: To 1'
STEMS: Erect, hairy
LEAVES: Compound, alternate, typically with five leaflets, hairy
FLOWERS: Blue, with white or pink
FRUITS: Legume, silky
FLOWERING: February to April
HABITAT: Frequent on clays, sands, and caliche soils in prairies, openings, and overgrazed pastures

Kairn's Sensitive-briar
Mimosa latidens (Small) B. L. Turner

Occurred in quail diets during the
winter, spring, and summer for crops
examined from South Texas.
Outline oval to almost circular.
Reddish to brown, somewhat
flattened, smooth, shiny, with a
faint U-shaped line on each face.
Length 3–3.7 mm; width 2.5–3 mm.

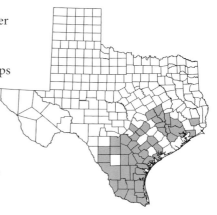

2 mm

Description of Plant

SYNONYM: Karne's sensitive briar
LIFEFORM: Native, perennial forb
HEIGHT: Low, trailing on ground
STEMS: Spreading or trailing, with curved prickles, to over 3' long
LEAVES: Compound, alternate, leaflets to just over ¼" long
FLOWERS: Pink, small, to ¾" across
FRUITS: Legume, covered with prickles, to 2⅜" long
FLOWERING: April to October
HABITAT: Frequent on sandy and loamy soils in prairies, dunes, disturbed areas, and openings

Gulf Indian Breadroot
Pediomelum rhombifolium (Torr. & A. Gray) Rydb.

Occurred in about 13% of 63 crops examined from South Texas during spring and comprised about 9.6% of total seed contents. Outline oval. Gray with black mottling, hilum white and on the margin, smooth, glossy, compressed. Length 2.8–3.2 mm; width 2.7–3 mm.

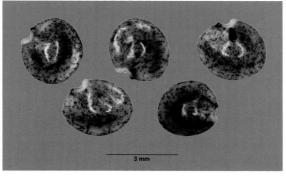

Description of Plant

SYNONYM: Roundleaf scurf-pea
LIFEFORM: Native, perennial forb
HEIGHT: To 6"
STEMS: Trailing, to 3' long
LEAVES: Compound, alternate, leaflets to about 1½" long
FLOWERS: Brownish-red, ¼" long
FRUITS: Legume
FLOWERING: May to June
HABITAT: Generally found on sandy soils, often among grass

American Snoutbean
Rhynchosia americana (Houst. ex Mill.)
M. C. Metz

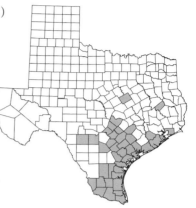

Occurred in about 39% of 108 crops examined from South Texas during summer and comprised about 4% of total seed contents. Outline oval to almost circular. Grayish with fine black spots, sometimes reddish-brown, flattened at margins, smooth, glossy. Length 3–4 mm; width 3–3.8 mm.

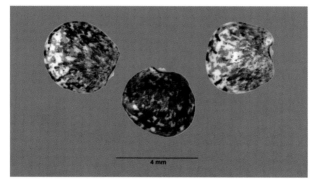

4 mm

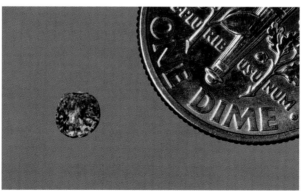

Description of Plant

LIFEFORM: Native, perennial vine or forb
HEIGHT: Less than 1'
STEMS: Trailing, hairy, angled
LEAVES: Simple, alternate, heart-shaped, hairy, leaves ½–5" long
FLOWERS: Yellow, to ⅜" long
FRUITS: Legume, flattened, to ¾" long
FLOWERING: March to November
HABITAT: Frequent on sandy and gravelly soils in openings, woodlands, and prairies

Prairie Snoutbean
Rhynchosia latifolia Nutt. ex Torr. & A. Gray

Comprised about 6% of total contents from crops examined from
North Texas. Outline somewhat
circular. Gray to reddish-brown
with black mottling, smooth,
glossy. Length 3.3–3.8 mm; width
3.2–3.5 mm.

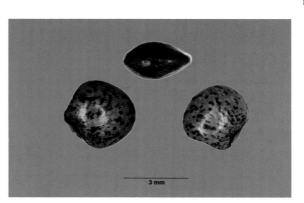

3 mm

Description of Plant

SYNONYM: Broadleaf snoutbean

LIFEFORM: Native, perennial vine or forb

HEIGHT: Low, usually trailing

STEMS: Twining, angled, with dense hairs, branching

LEAVES: Compound, alternate, three leaflets, hairy, diamond-shaped, to 2¾" long

FLOWERS: Yellow, five-petaled, to ½" long

FRUITS: Legume, flattened, hairy, with 1–2 seeds

FLOWERING: May to August

HABITAT: Frequent on sandy soils in woodlands and roadsides

Amberique-bean
Strophostyles helvola (L.) Elliott

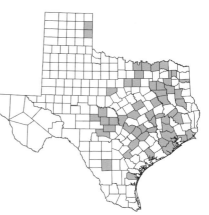

Occurred in about 38% of 16 crops examined from North Texas and comprised about 4.5% of total contents. Outline rectangular. Dark brown, hilum white, smooth, somewhat shiny. Length 6.4–7.8 mm; width 3.5–4 mm. Seeds contain about 22.3% protein.

7 mm

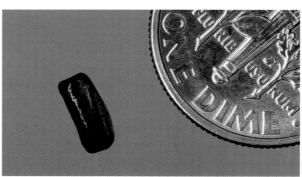

Description of Plant

SYNONYM: Trailing wildbean

LIFEFORM: Native, annual vine or forb

HEIGHT: Low-lying or climbing

STEMS: Trailing or twining, to 8' long

LEAVES: Compound, alternate, three leaflets

FLOWERS: White to light purple

FRUITS: Legume, hairy

FLOWERING: May to October

HABITAT: Found on sandy soils and shell deposits in grasslands, open woodlands, and near bay beaches

Slickseed Fuzzybean
Strophostyles leiosperma (Torr. & A. Gray) Piper

Occurred in 100% of five crops examined from North Texas and comprised about 35% of total contents. Outline rectangular with rounded corners. Light tan to reddish-brown, often mottled, hilum pale, smooth. Length 3.5–4.3 mm; width 2–2.8 mm.

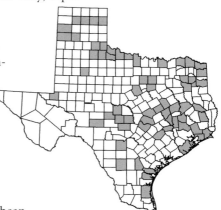

Description of Plant

SYNONYMS: Quail pea, Slickseed bean, Trailing wildbean
LIFEFORM: Native, annual vine or forb
HEIGHT: To about 20″
STEMS: Slender, trailing or twining, hairy on new growth
LEAVES: Compound, alternate, three leaflets, linear to elliptic
FLOWERS: Lavender to pink
FRUITS: Legume, slender, hairy, to about 1¼″ long
FLOWERING: May to September
HABITAT: Frequent on sandy or loamy soils in openings and along roadsides

4 mm

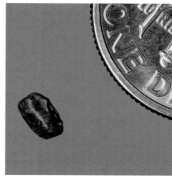

Pink Fuzzybean
Strophostyles umbellata (Muhl. ex. Willd.) Britton

Occurred in about 5% of 56 crops examined from East Texas during winter and comprised less than 1% of total contents. Outline rectangular to diamond-shaped with corners rounded. Gray to reddish-brown, often mottled, hilum pale, smooth. Length 3–4.2 mm; width 2.1–2.7 mm.

Description of Plant

SYNONYMS: Trailing fuzzybean, Perennial wildbean
LIFEFORM: Native, perennial vine or forb
HEIGHT: To about 18"
STEMS: Slender, trailing or twining, hairy on new growth
LEAVES: Compound, alternate, three leaflets, leaflets to about 1½" long
FLOWERS: Pink, on a long, leafless stalk
FRUITS: Legume, round, to about 2½" long
FLOWERING: June to September
HABITAT: Frequent on sandy soils in pine woodlands and roadsides

Leavenworth's Vetch

Vicia ludoviciana Nutt. ssp. *leavenworthii*
(Torr. & A. Gray) Lassetter & C. R. Gunn

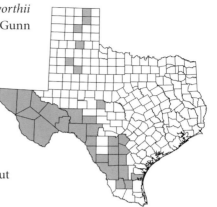

Occurred in about 37% of 67 crops
examined from South Texas during spring and comprised
about 35% of total seed contents.
Outline circular. Black to pale,
smooth, somewhat shiny, hilum
barely visible. Length 1.8–2.1 mm;
width 1.7–2 mm. Seeds contain about
33.2% protein.

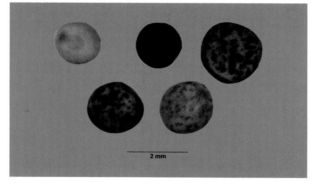

Description of Plant

LIFEFORM: Native, annual vine or forb

HEIGHT: To 2'

STEMS: Climbing or trailing, slightly hairy

LEAVES: Compound, alternate, 8–14 leaflets, leaflets up to ¾" long

FLOWERS: Violet to blue and white

FRUITS: Legume, flattened, to about ¾" long

FLOWERING: February to May

HABITAT: Frequent on a variety of soils in prairies, openings, and disturbed areas

PHOTO BY TIMOTHY E. FULBRIGHT

Viperina

Zornia bracteata Walter ex J. F. Gmel.

Occurred in quail diets during the summer and fall for crops examined from South Texas. Outline ovate. Red, flattened, smooth, somewhat shiny. Length 1.5 mm; width 1 mm.

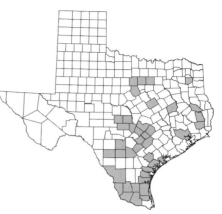

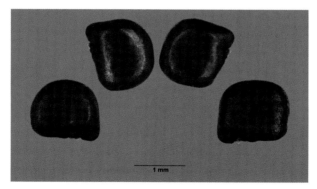

Description of Plant

SYNONYM: Bracted zornia
LIFEFORM: Native, annual forb
HEIGHT: To about 6"
STEMS: Trailing, hairy
LEAVES: Compound, alternate, smooth, four leaflets, leaflets to about
   ¾" long
FLOWERS: Yellow, with reddish-orange streaks
FRUITS: Legume, spiny, constricted around seeds
FLOWERING: April to June
HABITAT: Frequent on sandy or gravelly soils in openings and prairies

---

Carolina Geranium
Geranium carolinianum L.

Occurred in about 2% of 361 crops
examined from east-central
Texas and comprised less than
1% of total contents. Outline oval
to oblong. Dark red to brown or
grayish, surface with a network of
fine ridges. Length 2.2–2.3 mm;
width 1.4–1.6 mm. Seeds contain
about 22.9% protein.

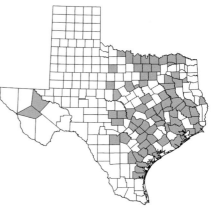

Description of Plant

SYNONYMS: Carolina cranesbill, Wild geranium, Stork's bill
LIFEFORM: Native, annual or biennial forb
HEIGHT: To 20"
STEMS: Erect or reclining, branched
LEAVES: Simple, alternate and opposite, with 5–9 lobes, to 2¾" long
FLOWERS: White to pink, five-petaled, about ¼" long
FRUITS: Capsule, one-seeded
FLOWERING: January to May
HABITAT: Frequent on sandy, rocky, and clayey soils in prairies, openings, woods, and disturbed areas

Bractless Blazingstar
Mentzelia nuda (Pursh) Torr. & A. Gray

One of seven forb species that
contributed to 40% of total
food volume from 963 crops
examined during fall and winter
from North Texas. Outline oval.
Tan, with a papery wing encircling the
seed, dull. Length 4 mm; width 2.5–3.5
mm. Seeds contain about 20.1% protein.

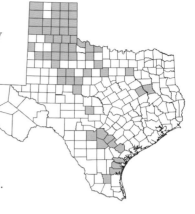

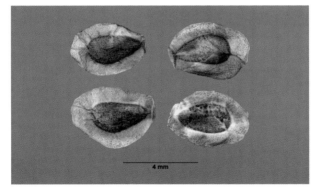

4 mm

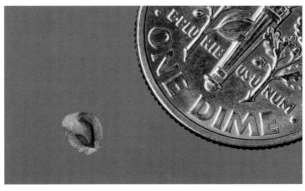

Description of Plant

SYNONYMS: Sand lily, Stickleaf mentzelia
LIFEFORM: Native, biennial or perennial forb
HEIGHT: To over 3'
STEMS: Erect, branched
LEAVES: Simple, alternate, rough, toothed, to 7" long
FLOWERS: White, showy, 4" wide
FRUITS: Capsule, cylindrical
FLOWERING: May to November
HABITAT: Occurs on deep sands in prairies and openings

MALVACEAE
Spreading Fanpetals
Sida abutifolia Mill.

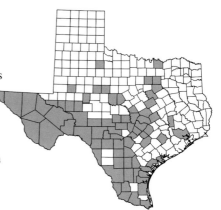

Occurred in about 69% of 91 crops examined from southwest Texas during fall and winter and comprised about 8% of total seed and fruit volume. Outline oval. Tan to black, smooth, dull. Length 1.5–2 mm; width 1–1.2 mm.

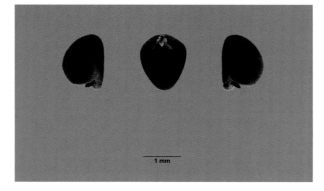

1 mm

Description of Plant

SYNONYM: Spreading sida
LIFEFORM: Introduced, annual or perennial forb
HEIGHT: To 18"
STEMS: Reclining to trailing, to 30" long
LEAVES: Simple, alternate, toothed, to about ¾" long
FLOWERS: Yellow, five-petaled
FRUITS: Capsule
FLOWERING: April to December
HABITAT: Frequent on a variety of soils in pastures, woodlands, and
open grassy areas

Low Menodora

Menodora heterophylla Moric. ex DC.

Occurred in 9% of 56 crops examined from East Texas during winter and comprised about 4.7% of total seed contents. Outline oval with a pointed tip. Tan, greenish, or brown, flattened, slight ridge on concave face, smooth, dull. Length 4–5 mm; width 2.5–3 mm.

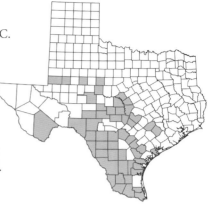

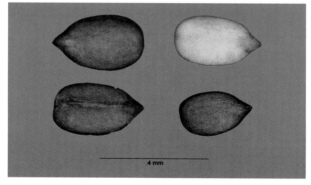

4 mm

Description of Plant

SYNONYMS: Redbud, Rough menodora
LIFEFORM: Native, perennial forb or subshrub
HEIGHT: To 10"
STEMS: Low-growing, with fine hairs
LEAVES: Simple, opposite, smooth, to 1" long
FLOWERS: Yellow, enclosed in red buds before bloom
FRUITS: Capsule, four-seeded
FLOWERING: February to December
HABITAT: Frequent on clays and heavy loams, caliche outcrops, hillsides, and in pastures

OXALIDACEAE
Slender Yellow Woodsorrel
Oxalis dillenii Jacq.

Occurred in about 82% of 67 crops examined from South Texas during spring and comprised about 11% of total seed contents. Outline elliptic, with one tip pointed. Reddish-brown with white ridges. Length 1.1–1.2 mm; width 0.8–0.9 mm.

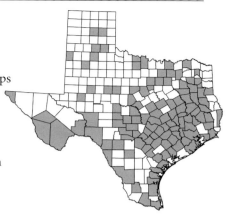

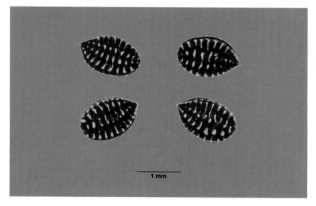

1 mm

Description of Plant

SYNONYMS: Dillens oxalis, Woodsorrel, Sourclover

LIFEFORM: Native, perennial forb

HEIGHT: To 16"

STEMS: Upright or reclining

LEAVES: Compound, alternate, clover-like, three leaflets, leaflets to ¾" long

FLOWERS: Yellow, five-petaled

FRUITS: Capsule, many-seeded

FLOWERING: March to May

HABITAT: Found on a variety of soils in prairies, openings, and disturbed areas

OXALIDACEAE

PAPAVERACEAE
Red Pricklypoppy
Argemone sanguinea Greene

Occurred in 19% of 63 crops
examined from South Texas
during spring and comprised
about 10% of total seed contents.
Outline circular with a pointed
tip. Black with rows of small pits,
shiny. Length 1.8–2 mm; width
1.5–1.7 mm. Seeds contain about
24.4% protein.

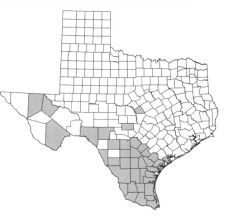

Description of Plant

SYNONYMS: Spiny pricklypoppy, Red poppy
LIFEFORM: Native, annual, biennial, or perennial forb
HEIGHT: To over 3'
STEMS: Erect, bluish-green, branching above, with prickles
LEAVES: Simple, alternate, prickle-margined, to 3" across
FLOWERS: White to purple
FRUITS: Capsule, spiny
FLOWERING: February to June
HABITAT: Common on sandy soils in prairies, openings, and disturbed areas, and along roadsides

Redseed Plantain
Plantago rhodosperma Decne.

Occurred in about 1% of 91 crops examined from southwest Texas during fall and winter and comprised about 1% of total seed and fruit volume. Outline oval. Light to dark red with a small white spot on concave face, smooth. Length 2.7–3 mm; width 1.3–1.6 mm. Seeds contain about 16.3% protein.

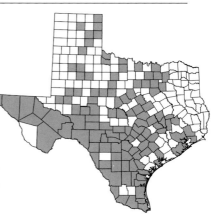

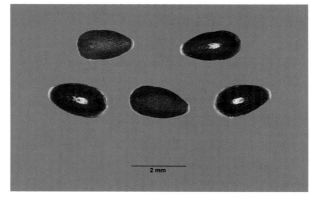

Description of Plant

SYNONYMS: Redseed indianwheat, Tallow weed
LIFEFORM: Native, annual forb
HEIGHT: To 14"
STEMS: Stemless or subterranean
LEAVES: Simple, hairy, growing at base of stem, to 6" long
FLOWERS: Whitish, small
FRUITS: Capsule, two-seeded
FLOWERING: March to May
HABITAT: Common on soils ranging from clayey to sandy in prairies and openings

Poorjoe
Diodia teres Walter

Occurred in about 9% of 56 crops
examined from East Texas
during winter and comprised
about 1% of total seed volume.
Outline oval to widely elliptic.
Brown to dark orange, rough, with
a ridge running lengthwise along
the seed. Length 3–3.6 mm; width
2.4–2.9 mm. Seeds contain about
13.6% protein.

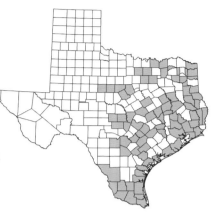

4 mm

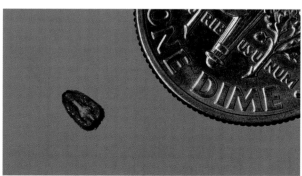

Description of Plant

SYNONYM: Rough buttonweed
LIFEFORM: Native, annual or perennial forb
HEIGHT: To 30"
STEMS: Branching, often purple, hairy
LEAVES: Simple, opposite, rough, linear, to about 1½" long
FLOWERS: White to pinkish
FRUITS: Capsule, usually two-seeded
FLOWERING: May to November
HABITAT: Frequent on sandy soils in prairies, fields, woods, and
    disturbed areas

SOLANACEAE
Starhair Groundcherry
Physalis cinerascens (Dunal) Hitchc.

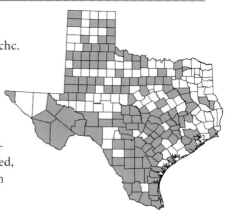

Occurred in about 57% of 63 crops examined from South Texas during spring and comprised about 15% of total seed contents. Outline oval to kidney-shaped. Tan to yellowish, flattened, smooth, somewhat shiny. Length 2.5–2.8 mm; width 2–2.2 mm.

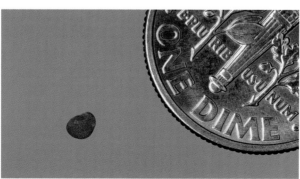

Description of Plant

SYNONYMS: Groundcherry, Smallflower groundcherry
LIFEFORM: Native, perennial forb
HEIGHT: To 1'
STEMS: Erect or drooping, covered with star-shaped hairs
LEAVES: Simple, alternate, triangular to oval, hairy underneath, to about 3" long
FLOWERS: Greenish-yellow, with spots
FRUITS: Berry
FLOWERING: February to December
HABITAT: Frequent on sandy soils in prairies, openings, and disturbed areas

Silverleaf Nightshade

Solanum elaeagnifolium Cav.

Occurred in about 3% of 33 crops examined from north-central Texas and comprised less than 1% of total seed volume. Outline usually rough oval, often variable. Yellowish-brown, smooth, thin with a small notch on the margin. Length 2.6–3.5 mm; width 1.9–2.5 mm. Seeds contain about 16.2% protein.

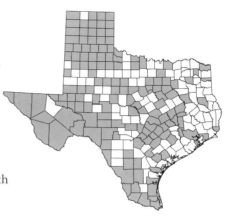

3 mm

Description of Plant

SYNONYMS: Tomatoweed, White horseweed, Trompillo
LIFEFORM: Native, perennial forb or subshrub
HEIGHT: To 30"
STEMS: Erect, with prickles, grayish
LEAVES: Simple, alternate, grayish-green
FLOWERS: Purple
FRUITS: Berry, yellow turning black
FLOWERING: April to November
HABITAT: Common on a variety of soils in prairies, openings, and disturbed areas

Buffalobur Nightshade
Solanum rostratum Dunal

Occurred in about 22% of 91 crops examined from north-central Texas and comprised about 16% of total seed volume. Outline variable with wavy margins. Dark brown to black with a network of pits over the surface. Length 2.4–2.7 mm; width 2–2.1 mm.

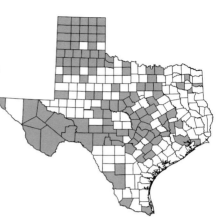

3 mm

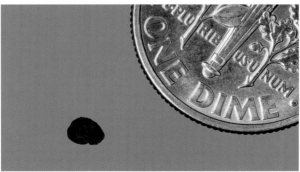

Description of Plant

SYNONYMS: Buffalobur, Kansas thistle

LIFEFORM: Native, annual forb

HEIGHT: To 2'

STEMS: Spreading, hairy, with long, slender spines

LEAVES: Simple, alternate, oval, with prickles, to 5" long

FLOWERS: Yellow, five-petaled

FRUITS: Berry with prickles

FLOWERING: June to October

HABITAT: Found on various soils in disturbed areas and along roadsides

ANACARDIACEAE
Eastern Poison Ivy
Toxicodendron radicans (L.) Kuntze

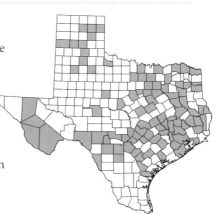

Occurred in about 9% of 56 crops examined from East Texas during winter and comprised about 1% of total seed volume. Outline figure 8–shaped. Tan to orange-red, smooth, surface lumpy. Length 5.1–5.8 mm; width 3.5–3.9 mm.

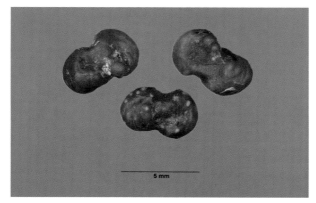

5 mm

Description of Plant

SYNONYM: Poison ivy
LIFEFORM: Native, perennial shrub or vine
HEIGHT: Low to very high-climbing
STEMS: Woody, hairy, to 4" in diameter
LEAVES: Compound, alternate, three leaflets, turning orange in fall, leaflets 1–8" long
FLOWERS: White to purple, five-petaled
FRUITS: Drupe, spherical
FLOWERING: April to May
HABITAT: Often abundant in woodlands and near streams

VITACEAE
Peppervine
Ampelopsis arborea (L.) Koehne

Occurred in about 9% of 33 crops
examined from north-central
Texas and comprised less than
1% of total seed volume. Outline
teardrop-shaped. Reddish-brown
to dark brown, shiny, raised reddish teardrop shape on rounded face,
two straight grooves on opposite face.
Length 4–4.7 mm; width 3.1–4 mm.

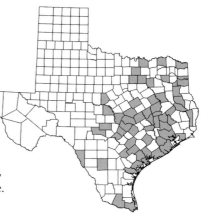

3 mm

Description of Plant

LIFEFORM: Native, perennial vine or shrub

HEIGHT: Low to moderately high-climbing

STEMS: Woody, bark light green to reddish, becoming tan

LEAVES: Compound, alternate, leaflets diamond-shaped, leaflets to 2" long

FLOWERS: Inconspicuous

FRUITS: Berry, green turning black

FLOWERING: June to August

HABITAT: Frequent on moist soils in woods, along streams, and sometimes in upland sites dominated by oaks

BERBERIDACEAE
Algerita
Mahonia trifoliolata (Moric.) Fedde

Occurred in about 8% of 67 crops
examined from South Texas
during spring and comprised
about 2.3% of total seed contents.
Outline oval to nearly elliptic.
Dark red to almost black, shiny.
Length 3–4 mm; width 1–2 mm.

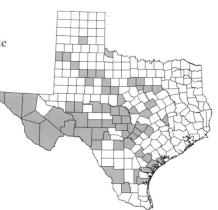

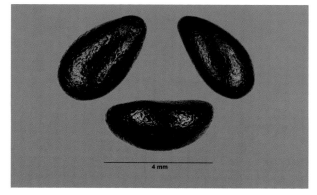

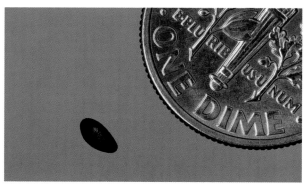

Description of Plant

SYNONYMS: Agarito, Desert holly, Chaparral berry
LIFEFORM: Native, perennial shrub
HEIGHT: To about 9'
STEMS: Bark gray, branching
LEAVES: Compound, alternate, three leaflets, leaflets tipped with spines
FLOWERS: Yellow, clustered
FRUITS: Berry, red
FLOWERING: February to March
HABITAT: Common on a variety of soils in openings and prairies

BORAGINACEAE
Woody Crinklemat
Tiquilia canescens (DC.) A. T. Richardson

Occurred in about 46% of 91 crops examined from southwest Texas during fall and winter and comprised about 4% of total seed and fruit volume. Outline oval to triangular. Gray to light brown, smooth, dull. Length 2–2.5 mm; width 2–2.2 mm.

2 mm

Description of Plant

SYNONYMS: Shrubby coldenia, Oreja de perro
LIFEFORM: Native, perennial subshrub
HEIGHT: To 1'
STEMS: Woody, ascending or spreading
LEAVES: Simple, alternate, hairy, to 6"
FLOWERS: White to pink to lavender
FRUITS: Nutlet
FLOWERING: March to November
HABITAT: Common on caliche ridges and limestone soils

BORAGINACEAE

Blackbrush Acacia
Acacia rigidula Benth.

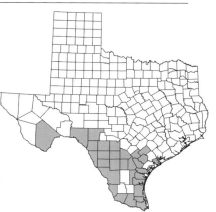

Occurred in 3% of 200 crops examined from South Texas and comprised about 1% of total seed contents. Outline oval to elliptic. Reddish-brown to black, compressed, pale U-shaped line on each face, smooth, somewhat shiny. Length 4–8 mm; width 2–3 mm.

6 mm

Description of Plant

SYNONYM: Chaparro prieto
LIFEFORM: Native, perennial, deciduous shrub or tree
HEIGHT: To 15'
STEMS: Multiple stems, with spines, bark whitish
LEAVES: Compound, alternate, leaflets to $^3/_5$" long
FLOWERS: White to light yellow
FRUITS: Legume, flattened, to $3\frac{1}{2}$" long
FLOWERING: February to July
HABITAT: Abundant on sandy, clayey, and gravelly soils, and often associated with guajillo on caliche ridges

Honey Mesquite
Prosopis glandulosa Torr.

Occurred in 28% of 200 crops examined from South Texas and comprised about 14% of total seed contents. Outline oblong, with one tip slightly pointed. Brown, flat, smooth. Length 6–7 mm; width 3.6–4 mm.

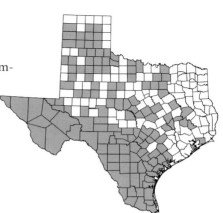

6 mm

Description of Plant

SYNONYMS: Mesquite, Texas mesquite
LIFEFORM: Native, perennial shrub or tree
HEIGHT: To 30'
STEMS: Bark rough, usually branching close to ground
LEAVES: Compound, alternate, 10–48 leaflets, leaflets to 2½" long
FLOWERS: White to cream-colored, cylindrical
FRUITS: Legume, brown, to 7" long
FLOWERING: March to September
HABITAT: Common on a variety of soils in brushlands, fields, and
   disturbed areas

Live Oak
Quercus virginiana Mill.

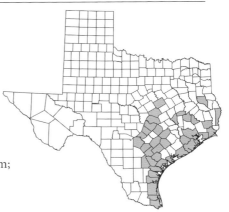

Occurred in about 4.4% of 565
crops examined from South
Texas and comprised about
5.5% of total contents. Outline oval. Tan and dark brown,
smooth, shiny. Length 20–27 mm;
width 11–13 mm.

20 mm

Description of Plant

SYNONYMS: Southern live oak, Texas live oak

LIFEFORM: Native, evergreen tree

HEIGHT: To about 60'

STEMS: Bark dark brown to black, deeply furrowed

LEAVES: Simple, alternate, thick, waxy above, to 3½" long

FLOWERS: Yellow

FRUITS: Acorns, to 1" long

FLOWERING: March to May

HABITAT: Frequent on sandy soils and well-drained clay loams, and along riverbeds, often forming shrubby thickets

---

Sweetgum
Liquidambar styraciflua L.

Occurred in about 1% of 185 crops examined from east-central Texas and comprised about 1% of total contents. Outline linear to oblong. Dark brown, usually with a tan papery wing attached, dull. Length 6–9.5 mm; width 1.8–2.3 mm. Seeds contain about 14.8% protein.

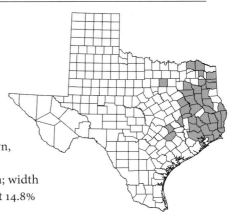

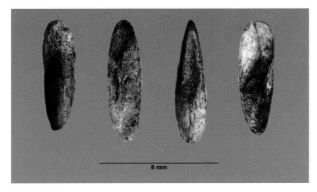

8 mm

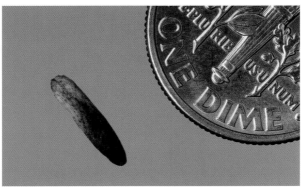

Description of Plant

SYNONYMS: Redgum, American sweetgum

LIFEFORM: Native, perennial tree

HEIGHT: To over 110'

STEMS: Bark gray to brown, furrowed stems, often corky

LEAVES: Simple, alternate, 3–7 lobes, to 7" long

FLOWERS: Small, solitary, green turning brown

FRUITS: Spherical, spiny, green turning brown, to 1½" across

FLOWERING: March to May

HABITAT: Found on sandy and clayey soils in low, wet areas and upland sites

MYRICACEAE
Wax Myrtle
Morella cerifera (L.) Small

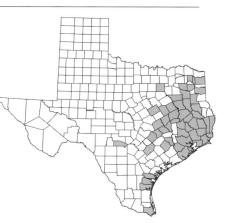

Occurred in about 7% of 167 crops examined from East Texas during winter and comprised about 4% of total seed contents. Outline oval. Tan to reddish-brown, rough. Length 3–3.7 mm; width 2.5–2.8 mm. Seeds contain about 4.9% protein.

3 mm

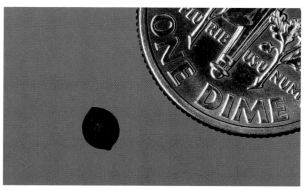

Description of Plant

SYNONYMS: Southern bayberry, Candleberry

LIFEFORM: Native, perennial subshrub, shrub, or small tree

HEIGHT: To 40'

STEMS: Bark gray, smooth, twigs slender and pale green

LEAVES: Simple, alternate, thin, evergreen, to just over 3" long

FLOWERS: Reddish, clustered

FRUITS: Drupe, clustered, white to bluish, waxy

FLOWERING: March to April

HABITAT: Found on moist soils in boggy grasslands, moist forests, and stream bottoms

---

MYRICACEAE

PINACEAE
Loblolly Pine
Pinus taeda. L.

Pinus spp. occurred in 52% of
167 crops examined from
East Texas during winter and
comprised about 31% of total
contents. *Pinus taeda* (pictured).
Outline triangular. Blackish, rough.
Length 5–6 mm; width 3.4–4.3 mm.
Seeds contain about 12.8% protein.

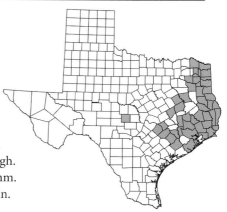

WOODY PLANTS, SHRUBS, AND TREES

Description of Plant

LIFEFORM: Native, perennial tree
HEIGHT: To 140'
STEMS: Bark dark brown, soft, thick
LEAVES: Needles in bundles of 2–3, green, to 9" long
FLOWERS: Yellow, clustered
FRUITS: Cone, scaly, to about 5" long
FLOWERING: July to August
HABITAT: Common on sandy, loamy, and gravelly soils in woods, hills, and savannahs

RUTACEAE
Hercules' Club
Zanthoxylum clava-herculis L.

Occurred in about 29% of 167 crops examined from East Texas during winter and comprised about 4.5% of total seed contents. Outline oval. Black with fine ridges, dull. Length 4–5 mm; width 3–3.5 mm.

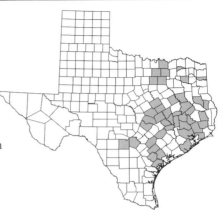

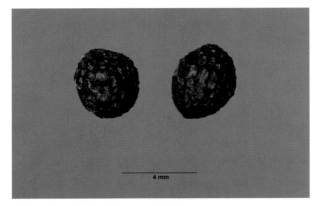

4 mm

Description of Plant

SYNONYMS: Pepperwood, Southern prickly ash
LIFEFORM: Native, perennial shrub or small tree
HEIGHT: To 13'
STEMS: With corky-based prickles, twigs brown to gray, stout
LEAVES: Compound, alternate, with 5–19 leaflets, stinging when chewed
FLOWERS: Greenish-white, four- or five-petaled
FRUITS: Follicle, brownish
FLOWERING: April to May
HABITAT: Occurs on clays and sands in woodlands, along ravines and streams

RUTACEAE

SAPOTACEAE
Gum Bully
Sideroxylon lanuginosum Michx.

Occurred in about 3% of 361 crops examined from east-central Texas and comprised about 1% of total contents. Outline oblong. Brownish, with light mottling, smooth, shiny. Length 9–9.5 mm; width 5 mm. Seeds contain about 8.7% protein.

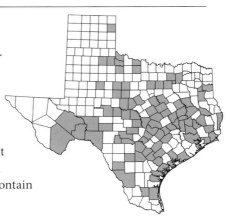

9 mm

WOODY PLANTS, SHRUBS, AND TREES

Description of Plant

SYNONYMS: Bumelia, Chittamwood, Woolybucket bumelia

LIFEFORM: Native, perennial shrub or tree

HEIGHT: To 60'

STEMS: Bark brown to gray, twigs zigzag

LEAVES: Simple, alternate or clustered, wedge-shaped, shiny, green, to 3" long

FLOWERS: White, five-petaled

FRUITS: Drupe, shiny, black

FLOWERING: May to July

HABITAT: There are four varieties of this species in Texas, growing on a variety of soils in pastures, woods, shell ridges, rocky outcrops, and along streams.

PHOTO BY MELODY LYTLE. COURTESY OF LADY BIRD JOHNSON WILDFLOWER CENTER

SAPOTACEAE

ULMACEAE
Spiny Hackberry
Celtis ehrenbergia (Klotzsch) Liebm.

Occurred in about 34% of 91 crops
examined from southwest
Texas during fall and winter and
comprised about 2% of total seed
and fruit volume. Outline circular
to broadly elliptic. Tan, compressed
at margins, tip slightly pointed, covered with fine ridges, dull. Length
4.5–5.5 mm; width 3–3.7 mm.

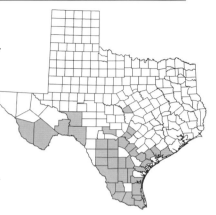

Description of Plant

SYNONYMS: Granjeno, Desert hackberry, Palo blanco
LIFEFORM: Native, perennial shrub or tree
HEIGHT: To 15'
STEMS: Bark smooth and gray, with paired spines, zigzag branches
LEAVES: Simple, alternate, toothed, green, to 2" long
FLOWERS: Small, greenish-white
FRUITS: Drupe, round, yellow to red, ½" across
FLOWERING: March to November
HABITAT: Found on a variety of soils in thickets, along fences, and
  under trees

Netleaf Hackberry

Celtis laevigata Willd. var. *reticulata* (Torr.) L. D. Benson

Comprised 5% of contents from crops examined from North Texas. Outline oval. Tan, nearly globose, tip slightly pointed, dull. Length 5.5–6 mm; width 4.5–5 mm. Seeds contain about 9.8% protein.

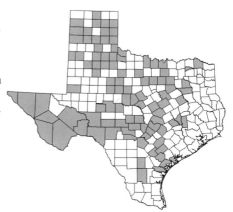

6 mm

Description of Plant

SYNONYM: Roughleaf hackberry

LIFEFORM: Native, perennial shrub or tree

HEIGHT: To 30'

STEMS: Bark gray, thin, warty, with light green to brown twigs

LEAVES: Simple, alternate, rough, prominently veined, to 3" long

FLOWERS: Inconspicuous, greenish

FRUITS: Drupe, green turning orange to brown

FLOWERING: March

HABITAT: Found on moist soils along streams and drainages and on canyon slopes and limestone hills, often in association with live oaks and mesquite

American Beautyberry
Callicarpa americana L.

Occurred in about 30% of 56 crops examined from East Texas during winter and comprised about 2.3% of total seed volume. Outline oval to elliptic. White to light brown, smooth, somewhat shiny. Length 2–2.9 mm; width 1.4–2 mm. Seeds contain about 6.2% protein.

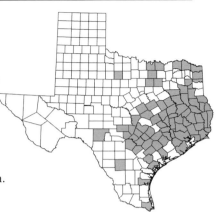

3 mm

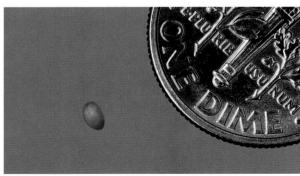

Description of Plant

LIFEFORM: Native, perennial shrub

HEIGHT: To 10'

STEMS: Woody, branching, brown to gray

LEAVES: Simple, opposite, margins toothed, wooly underneath, to about 6½" long

FLOWERS: Pinkish-white

FRUITS: Drupe, clustered, white turning purple

FLOWERING: June to July

HABITAT: Found on sandy soils in open woods and low areas

SUCCULENTS

CACTACEAE
Texas Pricklypear
Opuntia engelmannii Salm-Dyck ex Engelm. var. *lindheimeri*
(Engelm.) Parfitt & Pinkava

Occurred in 8% of 200 crops examined
from South Texas and comprised about
2.1% of total seed contents. Outline nearly circular with wavy
margins. Light tan with a darker
center, center of each flattened
face slightly raised, somewhat shiny.
Length 3.4–3.9 mm; width 3.1–3.8 mm.

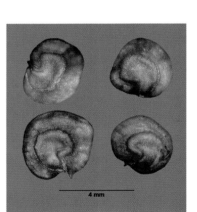

4 mm

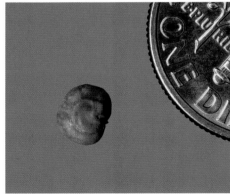

Description of Plant

LIFEFORM: Native, perennial succulent

HEIGHT: To 6'

STEMS: Erect or spreading, thick green pads with spines

LEAVES: None

FLOWERS: Yellow to red

FRUITS: Red, many-seeded

FLOWERING: April to May

HABITAT: Common on a variety of soils in thickets, openings, prairies, and coastal dunes

CHAPTER FOUR

Northern Bobwhite Habitat Management and Restoration

Habitat management for bobwhites involves manipulating vegetation to provide a set of food and cover resources that meet the annual life cycle needs of the bird. Maintaining these resources through management actions such as prescribed fire, soil disturbance, cattle grazing, and timber harvest provides the right configuration of food and cover that bobwhites require to persist through time. The concept of usable space—places on the landscape that bobwhites can use to meet their biological needs—is central to habitat management and restoration.

The purpose of this chapter is to outline the basic concepts of northern bobwhite habitat management. We begin with a brief history of the land uses that accidentally resulted in abundant and widespread populations of bobwhites. We then use the lessons learned from accidental management to lay the foundation for how to manage bobwhite habitat in a purposeful manner. Within this context, we then discuss how certain situations can exist where management may not work to increase bobwhite populations. We conclude with an overview of habitat restoration strategies that can be applied to places where bobwhite habitat has been lost.

Accidental Management

People have been actively managing habitat for northern bobwhites for nearly a century. However, the bobwhite habitat management prescriptions that we use today originated as an accidental by-product of land use during the early twentieth century. Millions of acres of bobwhite habitat were produced across the southeastern and midwestern United States as an accidental by-product of land use via row-crop farming and timber harvesting.

In the South, clearing forests for cotton and row-crop farming set the stage for a massive quail irruption at the turn of the twentieth century. The concept of plant succession—a progression of different types of plants that initially colonize a disturbed area and are then replaced by other plants as time progresses—is fundamental to understanding how land use changes influence bobwhite populations. Primitive farming techniques, combined with the collapse of the southern cotton industry after the Civil War, resulted in vast areas of early successional habitat dominated by small fields and weedy field borders that were interspersed with open canopy pine forests. Additionally, there was a culture and tradition of "woods burning" throughout the South, in which people intentionally lit fires to keep the forest understory open for travel access, as well as to discourage snakes, ticks, and other unwelcome organisms.

In the Central and Upper Midwest, a combination of primitive row-crop agriculture, cattle grazing of various levels of intensity, and widespread timber harvesting in the southern regions of the Lake States also set the stage for a massive increase in northern bobwhites in the nineteenth and early twentieth centuries. Again, the widespread disturbances from row-crop agriculture, a wide spectrum of grazing intensity, and timber harvesting resulted in a vegetation community dominated by a mixture of grasses, forbs, and occasional woody vegetation that provided excellent northern bobwhite habitat across a broad region of the country. Texas, because of its huge size and proximal location to both the South and the Midwest, also experienced elements of these same land uses that buoyed quail numbers from the early to the mid-1900s. All of these factors combined into a perfect storm that resulted in a massive increase in northern bobwhite populations on a huge scale.

Despite this remarkable set of circumstances, there were also large expanses of the landscape that did not support quail because they were overgrazed, burned either too frequently or too infrequently, or subjected to some other action that precluded emergence of habitat space for bob-whites. These areas were simply holes in the fabric of what was once widespread bobwhite habitat. Such places probably represented the first elements of fragmentation in the original quail wave that surged across the landscape of the eastern half of the continental United States.

During the past 50–100 years, depending on how you calibrate your starting point, bobwhite numbers have been slowly and steadily eroding across most of their geographic range in the United States. These trends are borne out by long-term data sets such as the Christmas Bird Count, USFWS Breeding Bird Survey, and numerous state wildlife agency surveys and monitoring programs. There are a few places that still sustain abundant numbers of wild quail in Texas (e.g., the South Texas Plains and the Rolling Plains), Oklahoma (the eastern portion of the Panhandle), parts of Kansas, and the Southeast (quail hunting plantations near the Thomasville and Albany, Georgia, areas). In the southeastern United States and East Texas, bobwhites persist in huntable numbers on public lands (primarily national forests, but also some national wildlife refuges) that are actively managed as open park-like pine forests with frequent prescribed fire for the endangered Red-cockaded Woodpecker. The common thread among all of these areas that still support relatively abundant bobwhite populations is that the people in charge of them practice purposeful management.

Purposeful Management

The first efforts at managing northern bobwhite habitat involved the use of prescribed fire in the southeastern United States. Early naturalists such as Val Lehmann, Herbert Stoddard, and Aldo Leopold realized that applications of prescribed fire every one to three years were required to maintain bobwhite habitat in the pine forests of the humid southeastern United States and the Pineywoods of East Texas. In the absence of fire, the pine forest understory quickly shifted from a plant community dominated by grasses and forbs—a habitat configuration that supported abundant populations of bobwhites—to one dominated by excessively thick and continuous woody shrubs and small midstory trees that did not produce or maintain bobwhite populations. As is the case today, these early naturalists understood the difference between applying fire under specific circumstances to promote a desirable habitat management outcome and the potential for habitat destruction by unchecked wildfires.

Over time, wildlife managers figured out how to maintain habitat for bobwhites by observing that different elements of the habitat were linked with the daily, seasonal, and annual cycle needs of the birds. In most of the current areas of Texas where quail remain, habitat management typically involves manipulating woody vegetation—primarily brush, although pines and hardwood trees are important in some of the eastern ecoregions—and herbaceous vegetation, which are grasses and forbs. Prescribed fire, while a useful bobwhite management tool on Texas rangelands, is not nearly as important in these regions as it is in the humid Southeast and the East Texas Pineywoods.

NESTING

Although bobwhites are capable of nesting almost anywhere, they prefer to locate their nests at the base of bunchgrasses that are at least two years old and about the size of a basketball (figure 4.1). Bunchgrasses, especially those in the genus *Andropogon*, provide excellent nesting cover when clumps are available at a minimum density of about 250 clumps per acre. It is important to note that these nesting clumps should be distributed across the landscape over many thousands of acres, and not merely located in isolated patches.

Establishing nesting cover for bobwhites where it does not exist can be a challenging prospect that can test the patience of any manager. This is because it usually takes several years for bunchgrasses to become established if they have been eliminated by overgrazing, incorrect applications of fire, or other factors. Typically, in these situations, the herbaceous plant community moves through a successional series of changes in which colonizing or ruderal species first appear. These are typically forbs such as croton, ragweed, sunflower, and other native broad-leaved flowering plants that, if left undisturbed, will eventually transition into vegetation where grasses, and eventually bunchgrasses, become more

FIGURE 4.1. An ideal bunchgrass clump for nesting. Illustration by Charles Wissinger

prevalent. While the value of these broad-leaved plants as nesting habitat is relatively low, they can potentially provide foraging habitat for broods and adults if located in a matrix of grass and brush, as we note below.

Maintaining nesting cover is a far easier management objective than establishing nesting cover where it does not exist. There are numerous ways of maintaining nesting cover through the use of fire, cattle, or mechanical management disturbances. Typically, if left undisturbed, many areas with bunchgrasses that once served as excellent nesting cover can eventually become too thick to be accessible to the birds. This point is probably reached when nest-clump density exceeds about 1,200 basketball-sized clumps per acre. When this range of bunchgrass density is reached, some form of management is needed to decrease the bunchgrass density and make the habitat once again usable by the birds. Grazing with stocker cattle until the average bunchgrass grass height is between 12–18 inches is one way of achieving this outcome. However, leaving too many cows on an area for too long a time will result in destruction of this critical nesting cover. If grazing is applied correctly in the context of a management prescription, the soil disturbance from cattle hooves also provides a microenvironment in and on the soil, which encourages growth of food-producing forbs around the clumps of remaining bunchgrass. Application of prescribed fire can also be used, although this typically results in a relatively quick (within a year or two) regrowth of dense bunchgrasses with few forbs. Following prescribed fire applications with discing during the months with frost may be a viable alternative to fire alone, or appropriate use of stocker cattle will also work. The point is that there are several different management tactics that can be used to achieve a strategy that results in good nesting cover for bobwhites.

BROODING AND FORAGING

Bobwhites eat seeds, insects, and green leafy vegetation. Everything that a bobwhite eats must be located either directly on the ground or within about six inches of the ground. Of course, some of the seeds eaten by bobwhites can be produced by relatively tall plants, such as sunflowers, woody shrubs, and even trees. However, it is critical to realize that the substrate from which bobwhites acquire these food resources is the ground. Furthermore, the dead leafy material, twigs, branches, and other obstructions such as dense grass, which all eventually end up on the ground, must be sufficiently sparse for the birds to both walk through an area and find their foods.

Imagine lying on the ground with the left or right cheek of your face on the ground and your vision peering down, up, or forward. This is about as close as you can get to understanding how bobwhites perceive their world. Now, add to this equation another two constraints: 1) you have a limited ability to scratch through the organic material on the surface of the ground to find the seeds that you need, and 2) you must constantly be on the lookout that some wily hawk or coyote is not pursuing

FIGURE 4.2. Desirable habitat architecture for bobwhites. Illustration by Charles Wissinger

you. The implication of these constraints is that bobwhites need a specific architectural configuration of habitat to meet their foraging needs. The vegetation structure must be open and accessible at ground level, but it should also contain overhead screening cover that provides them protection from aerial and ground predators (figure 4.2).

During the breeding season, bobwhite foraging habitat must provide arthropods (insects, spiders, caterpillars, etc.) in sufficient abundance for the newly hatched birds to grow and to meet the physiological needs of the nesting and breeding adult females. Young bobwhites eat only insects during their first two weeks of life because plant materials are insufficient to meet their physiological (i.e., growth) needs for protein. Adult female quail eat five times the amount of insects that males do during the breeding season (25% of their diet versus 5%, respectively), presumably because females produce eggs that are entirely protein except for the calcareous shell. Thus, it is critical that the habitat be structured in a manner that provides open patches at the ground level but have relatively dense overhead screening to both provide these resources and allow the birds to forage in relative safety from predators.

Areas that were mechanically disturbed by discing or hoof-action from cattle during the cool season months (November through March) often generate patches that are dominated by ragweed, croton, or sunflowers during the bobwhite breeding season. The phytophagous (plant-eating) insect communities associated with these plants provide the arthropod foods needed by the juvenile birds and adult females. An added benefit is that these same plants produce seeds relished by bobwhites during the winter months when they are in coveys.

LOAFING

Bobwhites typically feed twice during their daily cycle—once when they leave their nighttime roosts and again during late afternoon before they return to roost. During the midpart of the day, bobwhites typically seek cover in woody vegetation to avoid predators and digest the foods that they have accumulated in their crops. Thus, bobwhites need a place where they can spend a relatively large amount of the day in relative safety from their predators and avoid excess heat in summer. Typical bobwhite loafing cover consists of shrubs that are about 3–10 feet tall with associated canopy cover at least five feet in diameter. These areas can range from an isolated mesquite tree to a small motte. According to Dale Rollins, loafing sites should be distributed across the landscape at about the distance that an average adult can throw a softball, rather than clustered in specific locations.

In places where loafing sites are absent, a manager can encourage growth of shrubs and brush by allowing areas to grow in exclusion from fire, grazing, and other disturbances. However, in many places, the problem is not too little brush but too much brush. Thus, some of the basic concepts that pertain to brush management for bobwhites are applicable in this context. In the context of bobwhite management, brush management typically means using some form of mechanical method to reduce brush cover and provide conditions that are favorable for a mix of grasses and forbs that are distributed throughout the landscape.

Bobwhites have evolved in response to the constant threat of predation. Their explosive flight, cryptic coloration, and generally wary nature even when undisturbed all point to a behavioral arms race between them and their predators. Escape cover is one of the simplest elements of bobwhite habitat to provide and maintain. This is because escape cover is essentially nothing more than a patch of dense woody or herbaceous cover where a quail can either fly or run to in order to escape a predator when it is pursued. Almost all loafing cover can serve as escape cover, but small patches of escape cover are probably inadequate as loafing cover.

ROOSTING

During fall, winter, and early spring, when bobwhites are in coveys, they spend each night roosting in a circle with their tails pointing inward and their faces pointing outward. During periods of cold weather, the birds huddle tightly in a small circle to conserve heat. During warmer periods, these roost rings tend to be more loosely formed.

Typically, bobwhites roost at night in the middle of relatively large grassy areas that are as far from shrubs and mottes as possible. While this seems counterintuitive, it makes sense from the standpoint that if disturbed, each bird can quickly escape a predator by flying away in one of at least 10–12 directions without obstruction. In such cases, a landscape that has adequate bobwhite nesting cover will also most likely have adequate roosting cover.

The Landscape Perspective

The key to successful bobwhite habitat management is to provide the vegetation components in a configuration that allows them to meet their daily, seasonal, and annual cycle needs. This means that nesting, foraging, loafing, escape, and roosting cover should all be provided in relatively close proximity to each other (about 100–200 yards or less).

FIGURE 4.3. Excellent bobwhite habitat. Photo by Fidel Hernández

The result is a landscape with a patchwork mosaic of bunchgrasses, forbs, and woody vegetation that produces food, provides access to this food, and allows the birds at least a potential opportunity to escape predators (figure 4.3).

This also means that managers have to consider the scale at which they implement bobwhite habitat management. Obviously, it will be impossible to sustain bobwhites on an isolated 100–acre patch of habitat, no matter how optimal the configuration of the key features that quail need. Conversely, how to tackle and sustain habitat management when areas reach into the tens of thousands of acres presents a different set of challenges, such as how to allocate resources and prioritize management actions over such a broad area. Compared to having too little area to sustain quail, this is by far a preferable situation.

The concept of minimum habitat area is an issue that confounds opportunities for management success. The latest thinking is that for a bobwhite population to remain viable, or persist through time, it must consist of at least about 800 individuals. At a density of one bird per two acres, this means that at least 1,600 acres of habitat will be required. At lower population densities, even more acreage will be required.

One of the most vexing aspects of bobwhite habitat is that there is no single configuration of x% brush, y% grass and z% forb cover that will result in population persistence when a sufficient amount of habitat is available. Fred Guthery provided at least a partial solution to this problem by introducing the concept of "slack" in relation to bobwhite habitat components. The concept of slack basically states that bobwhites can persist and even do well in situations in which grass cover might be somewhat lacking, if sufficient brush cover is available to compensate for this lack of grass. Conversely, bobwhites can also survive and thrive in areas of relatively low brush cover (e.g., 10% or less) if sufficient grass and forb cover is available to compensate.

Cultural Management

People spend a great deal of money managing habitat for bobwhites. Unfortunately, too many times these funds are spent on cultural management actions rather than on purposeful management actions. The willingness of people to spend great sums of money on food plots, supplemental feed, and providing surface water stems from observations that make it seem like areas with these cultural improvements actually have more quail than places that do not have such improvements. The problem is that what is really happening is that these features attract and concentrate existing birds on an area, rather than actually produce more birds than the habitat could support without these cultural management features.

Fortunately, aspects of cultural management for bobwhites are largely neutral. That is, they probably do no harm, but they also probably do little good. The informed and effective wildlife manager realizes that the concepts we listed as purposeful management are the top priority for implementation. It is only after all the possible elements of purposeful management have been accomplished that aspects of cultural management should be considered. In other words, consider aspects of cultural management as the frosting that covers the purposeful management cake.

FOOD PLOTS

Planting agricultural crops such as sorghum, pearl millet, corn, and soybeans on a small amount of acreage (0.5–5 acres) is a popular quail habitat management practice in many places. Unfortunately, the payoff with respect to increased production of birds has not been proven through research, which points to the fact that this is a neutral practice. In Texas rangelands, the use of food plots is even more problematic because of frequent drought, which precludes their growth, or during years of adequate rainfall, which makes them unnecessary. In either case, the money spent on the tractor, seed, and fertilizer will be better spent on more purposeful aspects of management.

SUPPLEMENTAL FEED

People spend literally millions of dollars spreading corn, milo, soybeans, and other seed foods in the hopes that doing so will result in more quail. This is an unfortunate circumstance because hardly ever is food a factor that limits bobwhite populations, and numerous studies have shown that providing supplemental feed results in a neutral effect at best. The idea of distributing different forms of high-protein feed in pelleted form during the breeding season seems to be gaining popularity but needs to be evaluated through research.

PROVIDING WATER

Even in semiarid rangelands in Texas, bobwhites can meet their physiological needs for water, which is contained in vegetation. This so-called preformed water allows the birds to persist many miles from the available surface water at stock tanks, windmills, and other rangeland improvements for cattle.

Confusingly, observations of bobwhites frequenting locations with

---

available surface water appear to contradict the finding that they do not actually need this water to meet their daily needs. The logical explanation is that while bobwhites do not need surface water, per se, they still like to have access to it and thus find it attractive. Additionally, so-called observations of bobwhites "belly-dipping" their feathers into puddles and patches of surface water so that they can bring this moisture back to their nests for improved incubation lie in the realm of myth and legend rather than reality.

PREDATOR CONTROL

Many people think that if a manager implements a program of mammalian nest predator control, the result will be an increase in bobwhites. However, at least in many parts of Texas this is not likely the case. During the 1940s Val Lehmann observed that a massive coyote removal and reduction program on King Ranch did little or nothing to bolster quail numbers, even though research six decades later showed coyotes to be the top bobwhite nest predator in South Texas. Thirty years later, controlled predator removal field experiments by Sam Beasom and Fred Guthery supported Lehmann's observations. Essentially, what happens in South Texas is that the effects of rainfall, or lack thereof, combined with heat and habitat all work together in a manner that results in only modest, if any, increases in quail when mammalian nest predators are trapped. In other words, the effects of climate and habitat easily overwhelm any gains that might be achieved by nest predator reduction. This is somewhat different from the southeastern United States, where recent efforts at nest predator reduction appear to show a positive response in bobwhite production. Of course, shooting and removal of raptors—hawks and owls—is out of the question, as all of these species are federally protected.

When Management Does Not Work

Counterintuitive as it may seem, there are often situations in which management actions that are designed to stabilize and increase bobwhite populations end up doing neither. The causes behind situations in which management actions do not work to produce a desired population increase can be obvious, cryptic, or somewhere in between these two extremes. In most cases, for one reason or another, the cause behind why management actions have no impact is that they fail to eliminate some factor that is limiting the population. Most often, this happens in the context of cultural management actions. People often think that providing supplemental food or additional free water will result in a population increase. However, if neither food nor water is limiting, and often they are not, then these management actions will be neutral.

Applications of purposeful management can also have little potential impact on elevating or stabilizing quail populations as well. In these contexts, what usually happens is that management actions are either conducted at an inappropriate scale (e.g., too little soil disturbance, increases in nesting habitat that are insufficient) or over too small an area (e.g., less than 1,000 acres). Additionally, there are circumstances in which habitat, for one reason or another, is simply saturated with usable space, and no amount of additional management actions will increase the usable space for quail. Finally, there is the situation of weather. Long-term periods of drought and/or excess heat can easily overwhelm even the most comprehensive attempts to provide habitat resources for bobwhites through various management actions.

Habitat Restoration

Many millions of acres of bobwhite habitat have been lost across the geographic range of this species. The sources of these losses are diverse. They range from clean farming, high-density pine plantations, excessive grazing, suburban sprawl, and other vexing factors such as the widespread proliferation of exotic pasture grasses that provide little in the way of usable space for quail. In fact, most modern land uses are inherently hostile when it comes to providing habitat for bobwhites. This is the key reason that bobwhites are continuing to disappear across their geographic range.

It is critical to realize that any bobwhite habitat restoration effort will be context dependent. That is, how a restoration effort is conducted will depend on the set of circumstances that confronts a manager as well as the resources available to that manager.

One philosophical approach to habitat restoration for bobwhites on Texas rangelands is an overarching strategy to either add brush to huge areas of open grassy pastures (assuming most of the grasses are native species) or add areas of native grass pastures to huge areas of dense brush.

In either case, the basic idea is to identify limiting factors of the existing habitat and then implement management strategies to overcome these limitations. For example, excessive brush typically means inadequate grasses and forbs. Therefore, restoration actions that sculpt brush in a manner that allows loafing and escape cover to remain, while encouraging grasses and forbs to colonize the disturbed areas where brush was removed might be one strategy given such conditions.

Probably the most challenging bobwhite habitat restoration scenario is a situation in which the goal is to convert vast areas of exotic grasses back to usable space for quail. In many areas of Texas, cool-season exotic pasture grasses, such as bermudagrass, have been planted for the purpose of "improving" rangeland for cattle. The problem from a wildlife standpoint is that this is not an improvement at all. Rather, species such as bermudagrass completely eliminate usable space for quail. Converting pastures dominated by species such as bermudagrass to habitat for bobwhites requires repeated applications of herbicide to kill the grass, repeated applications of discing to kill the roots of the grass, and then probably several years for native vegetation (first forbs, then native grasses, then native brush) to colonize the areas where bermudagrass was removed. Unfortunately, there is not a simple cookbook recipe that can be followed by the numbers to solve these restoration challenges. Rather, each habitat restoration challenge, like each management situation, most likely requires a specifically tailored solution.

In Texas, the widespread presence of exotic grasses makes restoration of bobwhite habitat in these situations a huge challenge. The problem with most of these grasses is that they grow quickly and outcompete native grasses and forbs. Additionally, they often form a dense, thick mat of vegetation at the ground level that makes the habitat completely unusable for bobwhites. Recall the basic configuration of ideal nesting cover provided by native bunchgrasses and ideal brood and foraging cover provided by forbs. With species such as bermudagrass, the birds have no access to bare ground, and they have no screening cover overhead. This means they have no habitat. The habitat structure provided by exotic species such as bermudagrass and Old World bluestems such as yellow (King Ranch) and Kleberg's do not provide any of the critical elements of bobwhite nesting or foraging habitat structure, and as such produce little or no wild quail.

With respect to exotic grasses that tend to grow in bunches or clumps, the circumstances for bobwhites are somewhat different. For example, research has shown that buffelgrass can serve as a surrogate for native bunchgrasses as nesting cover, but the birds avoid it as foraging habitat. Also, areas dominated by buffelgrass tend to be able to produce only about half as many bobwhites as areas with native rangeland vegetation. Guineagrass is another exotic species that is sweeping across South Texas. While this giant panic grass may have the potential to provide food resources for bobwhites, we have only recently initiated research on how it influences bobwhite habitat use. All of the issues raised in this section on bobwhite habitat restoration need to be solved through research and management experiments if we are going to be able to return this bird to many areas of Texas where its habitat has been lost.

In Summary

It is impossible to provide much more than a thumbnail sketch of the basic elements of bobwhite habitat management in this short chapter. However, we hope that this essay has provided a portal that inspires you to dig further into the literature on management of bobwhites and other quails in Texas. Besides bobwhites, there are three other species of quails in Texas, two of which can be hunted (scaled quail and Gambel's quail) and one that cannot (Montezuma quail).

CHAPTER FIVE

Exotic Grasses

A Growing Problem for Northern Bobwhites

Texas has the most diverse grass flora of any state in the United States, with more than 500 native and exotic species. Perennial and annual grasses are important to northern bobwhites in virtually all phases of their life cycle. Perennial bunchgrasses, grasses that grow in distinct clumps (figure 5.1), provide critical nesting habitat (see chapter 4). Bobwhites eat seeds of certain species and young, growing green leaves of annual and perennial grasses, along with insects associated with the plants. Perennial bunchgrasses may also provide cover for thermoregulation and for hiding from predators.

Bobwhites use the base of perennial bunchgrasses to construct their nests. Sod-forming grasses form a dense, continuous turf undesirable for nesting (table 5.1). Bobwhites need about 25–75% of the ground surface to be relatively barren of vegetation and litter to facilitate movement and foraging for seeds and insects. The continuous turf formed by sod-forming grasses results in no bare ground for bobwhites to move around on. Planting turfgrasses, such as the exotic bermudagrass, eliminates the usable space of an area as bobwhite habitat.

Habitat for northern bobwhites is being lost to urbanization, cultivation, and other causes. Planting of exotic grasses, and invasion of these grasses into areas where they have not been planted, may result in reduced numbers of bobwhites in remaining habitat in Texas.

FIGURE 5.1. Perennial grass clumps, such as this clump of tanglehead, provide critical habitat for northern bobwhites to conceal their nests in. Photo by Timothy E. Fulbright

Exotic Grasses

The term "exotic" applies to grasses that did not originally occur in Texas, but were brought from a different continent or country and introduced to Texas by humans. Large-scale introduction of exotic grasses into North America resulted in part because drought and excessive grazing caused a decline in livestock numbers during the late-nineteenth and early-twentieth centuries. Ranchers attempting to survive these conditions during the first part of the twentieth century were hungry for productive grasses that would allow them to restore livestock numbers on depleted range- and pastureland. Government agencies and private organizations began to scour the world in search of "miracle" grasses adapted to the harsh growing conditions characteristic of western United States rangelands. Ecological ramifications of introducing grasses from other continents were not understood by scientists and agriculturalists involved in bringing these new plants to North America. By the end of the twentieth century, however, negative consequences of these introductions were becoming apparent to

FIGURE 5.2. Lehmann lovegrass. Photo by Timothy E. Fulbright

biologists and ecologists working in the rangeland ecosystems of the western United States.

Some plant species are highly invasive when introduced in new habitats where the competing plant species, natural vertebrate and invertebrate herbivores, and diseases with which they evolved are absent. Many of the exotic grasses introduced to Texas and the southwestern United States persisted and reproduced where they were planted and spread rapidly into the surrounding landscape. In some cases, exotic grasses spread and replaced native grasses and forbs in a relatively short period of time.

Large-scale planting of grasses such as Lehmann lovegrass and buffelgrass provided point sources for their spread to areas where they were not planted (figures 5.2 and 5.3). Seeds of both species can travel far from established plants by clinging to people, animals, and vehicles. The area covered by Lehmann lovegrass in Arizona, for example, more than doubled between 1950 and 1984 as a result of the grass spreading beyond areas where it was initially seeded. Invasion of these grasses is ongoing. For example, buffelgrass was planted on more than 10 million acres of southern Texas and 16 million acres of Mexico between 1949 and 1985. According to recent estimates, buffelgrass will eventually invade 12% of the entire country of Mexico because of its ability to spread aggressively and replace native plants. The extent to which these grasses continue to spread and replace native plant communities in Texas has not been documented.

FIGURE 5.3. Buffelgrass. Photo by Timothy E. Fulbright

FIGURE 5.4. Kleberg's bluestem. Photo by Timothy E. Fulbright

Future climate changes predicted for Texas may further exacerbate the increase in exotic grasses. Temperature-sensitive subtropical grasses, such as buffelgrass and guineagrass, will likely extend their range northward in latitude as minimum temperatures increase. Summer precipitation has increased in portions of southern Texas, which may favor further increases in exotic grasses adapted for growth during hotter portions of the year.

Many of the exotic grasses in Texas have attributes that are undesirable for bobwhites. Exotic grasses that negatively impact habitat occupied by northern bobwhites include bahiagrass, bermudagrass, buffelgrass, Lehmann lovegrass, and Old World bluestems including yellow and Kleberg's bluestem (table 5.2) (figure 5.4). Tall fescue is also highly detrimental to bobwhite habitat, but it is not widespread in Texas. Other exotic grasses abundant in Texas include guineagrass, Johnsongrass, Kleingrass, Rhodes grass, weeping lovegrass, Wilman lovegrass, and rose Natal grass. Effects of the latter group of grasses on bobwhite populations are unclear.

Bahiagrass is native to Mexico and Central and South America and has been extensively planted as a pasture grass in the southeastern United States. The grass tends to form dense monocultures and excludes other grasses and forbs. In Texas, bahiagrass occurs primarily in the eastern portion of the state and along the coast. Although bobwhites eat the seeds, the dense canopy formed by bahiagrass is highly detrimental to bobwhite habitat because it reduces bare ground below the threshold needed for movement and foraging.

Bermudagrass probably originated in Asia and is the one of the most important pasture grasses in the southeastern United States. Coastal bermudagrass has been planted on at least 10 million acres in the southern United States as cattle forage; other varieties of bermudagrass have also been widely planted. Conversion of native grass pastures to bermudagrass is the major factor in the near disappearance of bobwhites from the Blackland Prairies and Post Oak Savannah regions of Texas. In spite of the degradation of wildlife habitat the grass causes, it is still highly regarded for planting as livestock forage.

Buffelgrass is native to India and Africa and occurs in western and southern Texas. Buffelgrass predominates and spreads where summer rainfall ranges from 6 to 20 inches, winter rainfall is less than 16 inches, and winter temperatures are seldom less than 41°F. The grass spreads and becomes dominant on loamy textured soils that are relatively low in nitrogen and does not persist in soils that are high in nitrogen. Buffelgrass is a valuable forage grass for domestic livestock and is held in high regard by many livestock producers because of its productivity, drought tolerance, and palatability to livestock.

Buffelgrass is rhizomatous and can outcompete other plant species for soil moisture, making it an aggressive and highly successful invader in semiarid rangelands. In a research project in southern Texas, buffelgrass spread outside areas where it was planted and replaced the surrounding native plant community with a buffelgrass monoculture in only three years.

Buffelgrass may persist in scattered clumps, along roads, or at the canopy fringe of mesquites in areas where it is only marginally adapted to the climate or soils. The grass does not appear to be detrimental to northern bobwhites in these areas. Bobwhites readily nest in tall clumps of buffelgrass that have not been excessively grazed by domestic livestock. Areas dominated by buffelgrass are much poorer habitat for bobwhites, however, than areas dominated by native grasses. Plant communities dominated by buffelgrass are lower in native plant species diversity, support fewer forbs, and have fewer forage arthropods than communities dominated by native grasses. In a recent study in southern Texas, sites dominated by buffelgrass and Lehman lovegrass supported half the bobwhite population densities compared to what was found on sites dominated by native grasses. Foraging bobwhites avoid areas where buffelgrass canopy cover exceeds 30%, the birds preferring to forage in areas where forbs are more abundant and diverse.

Lehmann lovegrass originated in tropical and subtropical South Africa. The grass predominates and spreads in areas where rainfall during active growth exceeds six inches and daily mean minimum temperatures are 25–68°F. It grows best on sandy and sandy loam soil. Lehmann lovegrass produces an abundance of seeds that germinate rapidly, enabling it to reestablish more successfully than natives following fire and drought. Drought or fire can therefore result in the replacement of native grasses communities by Lehmann lovegrass monocultures.

Landscapes dominated by Lehmann lovegrass support lower abundance of northern bobwhites and other bird species than landscapes dominated by native plants. In a study in the desert grasslands of Arizona, breeding bird diversity was lower in Lehmann lovegrass–dominated landscapes than in native grassland.

Yellow and Kleberg's bluestems are included in a group of grasses referred to as Old World bluestems, which originated in North Africa and parts of Asia and Europe. Old World bluestems have been planted throughout Texas for livestock forage and erosion control. About four million acres of cropland in the USDA Conservation Reserve Program have been planted with Old World bluestems in North Texas. Yellow and Kleberg's bluestems tend to establish along roads, possibly because of the dissemination of seeds caught on vehicles. Seeds of these grasses are not desirable quail foods. Direct effects of invasion or planting of

FIGURE 5.5. Guineagrass. Photo by Timothy E. Fulbright

FIGURE 5.6. Rose Natal grass. Photo by Forrest Smith

these grasses on bobwhite habitat have not been researched, but they form dense monocultures that reduce forb diversity and abundance. Old World bluestem pastures in Kansas support fewer bird species and more than four times less arthropod biomass than pastures dominated by native grasses. Lower arthropod biomass in the Old World bluestem pastures may have resulted from lower canopy cover of forbs in the Old World bluestem pastures than in the native grass pastures.

Guineagrass (figure 5.5) may have accidentally been introduced to North America from slave ships as early as 1684. The grass originated in tropical Africa and the subtropics of South Africa. Guineagrass is invasive, but it is unclear whether or not its presence affects bobwhite habitat quality. Guineagrass is a bunchgrass, and the seeds are as readily eaten, or more readily eaten, by bobwhites than several native grasses, based on preference trials conducted in our aviary. Seed size is so small, however, that guineagrass is of little value as a bobwhite food plant. Dense stands of guineagrass greatly suppress forb abundance, which in turns restricts food availability for bobwhites.

Rose Natal grass (figure 5.6) originated in tropical Africa and subtropical South Africa. The grass is invasive, but it is a bunchgrass and does not tend to form dense monocultures. Thus, rose Natal grass may be rather innocuous as a component of quail habitat.

Johnsongrass (figure 5.7) originated in the Mediterranean region and North Africa. Johnsongrass seeds are eaten by bobwhites and are large enough to be of some value as food, but the grass tends to form dense stands when it is not grazed by livestock, which reduces plant species diversity and results in a configuration similar to many of the other sod-forming exotic species.

Kleingrass, blue panicum, Rhodes grass, weeping lovegrass, and Wilman lovegrass tend to be less invasive than buffelgrass and Lehmann lovegrass and tend to remain where they were planted. Kleingrass originated in tropical Africa and subtropical South Africa. Claims that Kleingrass seeds are consumed by bobwhites exist in the literature, but there is no published evidence that we are aware of that documents

FIGURE 5.7. Johnsongrass on former cropland in Central Texas. Photo by Timothy E. Fulbright

FIGURE 5.8. A near monoculture of Kleingrass on former cropland in Central Texas. Photo by Timothy E. Fulbright

that seeds of the grass constitute an important part of bobwhite diets. The seeds are so small that a bobwhite would probably expend more energy than it gained if it spent a large amount of time feeding on Kleingrass seeds. If a bobwhite lived in a Kleingrass field where seeds of other plants were not available, the bird would have to eat 32,870 Kleingrass seeds per day to meet its caloric needs. In contrast, a bobwhite can meet its energy needs with only 1,160 common sunflower seeds. "Verde" Kleingrass is a variety of the grass with larger seeds than other Kleingrass varieties and is purported to be better food for quail. A good stand of native forbs, however, provides far better food for bobwhites than a monoculture of Kleingrass (figure 5.8).

Blue panicum is often recommended as a good seed-producing plant for bobwhites. Evidence of the value of this grass to bobwhites as food is lacking. Weeping lovegrass pastures support lower arthropod diversity and abundance than pastures dominated by native grasses. Otherwise, information on effects of weeping lovegrass and Wilman lovegrass on bobwhite habitat is also lacking. These grasses are not as likely as buffelgrass or guineagrass to invade outside of areas where they have been planted, and thus may not impact habitat on the scale that many of the other exotic grasses do.

Effects of Exotic Grasses on Bobwhite Habitat

Possible factors causing the decrease in abundance of bobwhites when exotic grasses are planted or invade include reduction in food availability or a reduction in usable space. Diversity and abundance of native grasses and forbs that provide seeds eaten by bobwhites often decline when exotic grasses are planted or invade. Forb diversity and abundance decline with increasing canopy cover of buffelgrass, bermudagrass, guineagrass, Lehmann lovegrass, and Old World bluestems. Reduced abundance of seed-producing native grasses and forbs may result in lower carrying capacity of the habitat for bobwhites. Arthropods eaten by bobwhites are also less abundant in habitats dominated by exotic grasses. Arthropods are important because they are calorie rich compared to individual seeds and provide high levels of protein needed by laying hens and growing chicks.

Many exotic grasses, particularly sod grasses such as bermudagrass, form dense monocultures that may restrict bobwhite movements or inhibit foraging activity, resulting in avoidance of these areas by the birds. Bermudagrass pastures are avoided by bobwhites and should not be considered to be bobwhite habitat by managers making decisions about the number of birds an area should support (figure 5.9).

Short, sod-forming grasses such as bermudagrass provide poor nesting habitat for northern bobwhites. Bunchgrasses such as buffelgrass are used by bobwhites for nesting. In a study in South Texas, bobwhites nested in buffelgrass pastures within 100 yards from the edge of brush-dominated rangeland, but they avoided the interior of the fields for nesting.

Managing Exotic Grasses

REDUCING SPREAD OF EXOTIC GRASSES

Exotic grasses continue to be widely planted in Texas for livestock forage and erosion control. Exotic grasses are advertised by seed companies and are sometimes included in seed mixes for wildlife. Widespread planting of these grasses in pastures, roadsides, and Conservation

FIGURE 5.9. A bermudagrass pasture. Photo by Timothy E. Fulbright

Reserve Program plantings increases the probability that areas that have not been invaded by these grasses will be in the future. Invasion of exotic grasses is difficult to prevent where soil physical and chemical properties and climate are conducive to their establishment. There are, however, ways to inhibit their establishment or prevent exotic grasses from forming monocultures.

Native grasses and forbs should always be used in all revegetation and rangeland seeding projects if providing good bobwhite habitat is a management objective. This includes plantings along roadsides and pipeline rights-of-way, revegetation of abandoned oil field sites, and seeding following brush management. One reason that seed dealers often recommend planting exotic grasses such as buffelgrass is that they are more productive and easier to establish than natives. You must remember, however, that when you plant exotics you are trading off habitat quality for production of livestock forage and ease of establishment. By planting invasives such as bermudagrass, buffelgrass, Lehmann lovegrass, and others, you are providing a foothold for these grasses to spread and replace native plant communities. Local seed dealers or representatives of agencies such as the Texas Parks and Wildlife Department and Natural Resources Conservation Service can provide advice on locally adapted native plant alternatives to exotics.

Rangeland discing is a common practice used to disturb soil, promote growth of early succession weeds for bobwhites, and create travel lanes for quail-hunting vehicles. Disturbing soils where environmental conditions are conducive to establishment of exotic plants can potentially result in conversion of native plant communities to monocultures of exotic grasses. Discing and other forms of soil disturbance should be avoided where scattered individuals of exotic grasses are already present. Presence of clumps of these grasses indicates that the soil seed bank contains their seeds. In a research project conducted in Starr County, Texas, discing once in an area where sparse, scattered plants of buffelgrass were present resulted in an exponential increase in the exotic grass (figure 5.10). The dramatic increase in buffelgrass following discing was not immediate; a significant cover of buffelgrass did not develop until five years following the initial disturbance. Johnsongrass may also be increased by discing, particularly where livestock are not present. Alternatives to discing to increase early successional forbs include livestock grazing. Cattle grazing, however, causes some soil disturbance and may indirectly result in an increase in exotic grasses if the animals disseminate exotic grass seeds.

Exotic grasses are not invasive on all soils. In the Starr County study, buffelgrass increased on Ramadero loams but not on Delmita fine sandy loams. It is critical for land managers to develop knowledge of what soils are susceptible to invasion of exotics and what soils are not. In general, soils along drainages, streams, and rivers tend to be more susceptible to invasion and development of exotic grass monocultures than dry, infertile, upland soils.

Mechanical brush clearing using methods such as roller chopping, aerating, and root plowing also cause soil disturbance and may result in an increase in exotic grasses. Roller chopping creates less soil disturbance than root plowing and raking, which remove the native plant community. Likelihood of invasion of exotic grasses and formation of monocultures is therefore much greater following root plowing than it is after less severe disturbances such as roller chopping or mowing. As with discing,

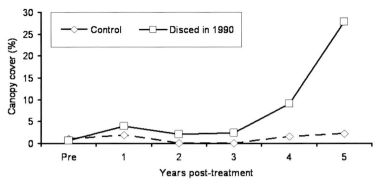

FIGURE 5.10. Effects of discing on buffelgrass on Ramadero loam soils. Adapted from Johnson and Fulbright (2008)

proximity of established exotic grasses; soil physical and chemical properties; and environmental conditions, particularly rainfall, are all important factors determining susceptibility to invasion by exotic grasses.

Prescribed burning is a widely used habitat management tool for northern bobwhites (see chapter 4). Many of the African grasses that have been introduced into the rangelands of Texas, including buffelgrass and Lehmann lovegrass, are stimulated by fire. Prescribed burning should not be used in areas prone to invasion by these grasses. Fire may convert stands of native grasses to stands of Lehmann lovegrass because Lehmann lovegrass recovers more rapidly than native grasses following fire. In addition, high seedling emergence following fire allows Lehmann lovegrass to outcompete and replace native grasses.

Buffelgrass may increase following prescribed burning, particularly on soils where it is well adapted. Prescribed burning at two- to three-year intervals maintains maximum productivity of buffelgrass stands; stands that are not burned decline in productivity. Buffelgrass produces seeds following fire even under dry growing conditions. Productivity of buffelgrass stands may decline if summer burns are followed by a dry spell, which increases damage to buffelgrass stands by desert termites.

Roads may act as conduits for the spread of exotic grasses. Seeds of grasses such as Old World bluestems and buffelgrass adhere easily to vehicles and are transported long distances from areas where the grasses are abundant. In the Edwards Plateau region of Texas, yellow bluestem is more abundant near roads than in other parts of the landscape because the roads facilitate dispersal of the seeds. Cleaning seeds from vehicles and equipment entering ranches and other areas that are free of exotic grasses may help to reduce their spread.

Grazing by domestic livestock does not appear to directly cause an increase in exotic grasses. Grazing, however, may contribute to invasion of exotic grasses indirectly. Lehmann lovegrass may expand from seeded areas through seed dissemination by domestic livestock. Lehmann lovegrass seeds are not destroyed in the digestive tract of sheep and cattle and may germinate or establish in dung or following the decay of dung. Ingested seeds may remain in the digestive tract of livestock for several days. Animals that travel or are transported away from areas where Lehmann lovegrass is present may disseminate the seeds into non-invaded areas.

MANAGING LANDSCAPES ALREADY DOMINATED BY EXOTIC GRASSES

Minimizing disturbance is of little value if the landscape is already dominated by exotic grasses. In situations where exotics have replaced the native vegetation, soil disturbance may provide some benefits, although they will only be temporary. Canopy cover of guineagrass in dense stands of the plant may reach nearly 100%. In a recent study in South Texas, canopy cover of forbs in dense stands of guineagrass was only 7% during May. Spraying 1.3 pounds active ingredient per acre glyphosate followed by tilling reduced guineagrass canopy cover and resulted in an almost six-fold increase in forb canopy cover. The effects were only short-lived, however, because in only six months guineagrass had returned to more than half of its pretreatment cover. You would need to spray strips or patches within a pasture with glyphosate every one to two years to maintain increased forb abundance and diversity.

Grazing of dense stands of buffelgrass by cattle may benefit bobwhites. Bobwhites use ungrazed buffelgrass clumps near the edges of dense stands of buffelgrass for nesting, but the interior of dense stands of ungrazed buffelgrass is avoided. Feeding, brooding, and travel areas can be created by grazing patches of buffelgrass to reduce canopy cover and increase bare ground. A mosaic of grazed patches of buffelgrass interspersed with ungrazed patches for nesting should theoretically provide better bobwhite habitat than a monoculture of dense buffelgrass because feeding areas and access to nest sites are created.

Areas dominated by bermudagrass or tall fescue can be successfully restored to native plant communities. Guidelines for converting bermudagrass pastures to bobwhite habitat (recommended by Texas Cooperative Extension) are as follows:

- Burn, mow, or graze heavily during late winter to stimulate active growth the following spring.
- The following spring when bermudagrass is growing and exceeds six inches tall, apply 41% active ingredient glyphosate; use 4 quarts per acre on sandy soils and 6 quarts per acre on clay soils.
- If the bermudagrass sod is not completely killed, repeat the herbicide treatment.
- About two weeks after the sod is completely dead, drill native grass and forb seeds into the dead sod.
- In areas with higher rainfall, you can disc the dead bermudagrass sod and broadcast native grass and forb seeds.
- If repeated herbicide applications are needed during the summer, planting should be delayed until autumn.

Guidelines for converting tall fescue pastures to bobwhite habitat (developed by Dr. Thomas G. Barnes and his graduate students at the University of Kentucky) are as follows:

- Burn the pasture in late winter.
- When the grass is six to eight inches tall in spring, apply imazapic at 0.18 pounds active ingredient per acre with 0.2 gallons per acre methylated seed oil and 0.2 gallons per acre 28:0:0 liquid fertilizer.
- Drill native grass and forb seeds about one month following herbicide application.

Converting pastures dominated by buffelgrass, guineagrass, Old World bluestems, and other exotics to bobwhite habitat is more difficult than converting bermudagrass or tall fescue pastures. Application of glyphosate and imazapic reduces buffelgrass canopy cover for less than a year, and the exotic soon chokes out planted native grasses and forbs. Applying tebuthiuron at 2 pounds per acre reduces buffelgrass canopy cover for two years, but the treatment also kills forbs and does not reduce buffelgrass sufficiently to allow establishment of native grasses.

Applying glyphosate during early stages of vegetative growth and then again eight weeks later dramatically reduces Caucasian bluestem. A single herbicide application does not work well because dormant buds are located at the base of the plant. Killing growing tissues with herbicide stimulates growth of the basal buds, and the plant simply replaces the stems that were killed by the herbicide. Treating the plant twice kills the stems produced from the dormant basal buds. A problem with this approach is that glyphosate also kills native vegetation. Old World bluestems often elevate stems and seedheads before native grasses. A possible way to make the herbicide treatment more selective is to use a ropewick that coats the taller stems and seedheads of the Old World bluestems before the native grasses have elevated their seedheads. Clearly, much additional research is needed to develop more effective protocols for controlling warm-season exotic grasses.

TABLE 5.1. Common Texas Grasses. Grasses eaten by northern bobwhites or possessing characteristics making them potential nesting cover are marked with an X. Grasses that form dense stands that potentially may restrict movements or foraging are also marked.

Common name	Seeds eaten	May provide nesting cover	May restrict movement or foraging
Alkali sacaton		X	
Arizona cottontop		X	
Bahiagrass	X		X
Barnyardgrass	X		
Beaked panicgrass	X		
Bermudagrass			X
Big bluestem		X	
Big sandbur		X	
Black grama		X	

TABLE 5.1. Continued.

Common name	Seeds eaten	May provide nesting cover	May restrict movement or foraging
Blue grama			
Blue panicum	X	X	
Broomsedge bluestem		X	
Brownseed paspalum	X	X	
Buffalograss			
Buffelgrass		X	X
Burrograss			
Bush muhly			
Canada wildrye			
Cane bluestem		X	
Common carpetgrass			
Composite dropseed		X	
Curly-mesquite			
Cylinder jointtail			
Dallisgrass	X		
Eastern gamagrass		X	
Fall witchgrass	X		
Florida paspalum	X	X	
Green sprangletop		X	
Gulf cordgrass		X	
Hairy grama			
Hairy woollygrass			
Hall's panicgrass	X		
Heller's rosette grass	X		
Hooded windmill grass		X	
Indiangrass		X	
Johnsongrass	X		X
Kleberg's bluestem			X
Kleingrass	X	X	
Little barley	X		
Little bluestem		X	
Longtom			
Marsh bristlegrass	X		
Multiflower false Rhodes grass		X	
Perennial ryegrass			

TABLE 5.1. Continued.

Common name	Seeds eaten	May provide nesting cover	May restrict movement or foraging
Pink pappusgrass		X	
Plains lovegrass			
Prairie threeawn			
Purple lovegrass		X	
Purple threeawn			
Purpletop tridens		X	
Red grama			
Red lovegrass			
Rescuegrass	X		
Saltgrass			
Saltmeadow cordgrass			X
Sand dropseed			
Sand lovegrass		X	
Sideoats grama		X	
Silver beardgrass		X	
Smut grass			
Splitbeard bluestem		X	
Streambed bristlegrass	X	X	
Switchgrass	X	X	
Tall fescue		X	X
Tanglehead		X	X
Texas bluegrass		X	
Texas cupgrass	X	X	
Texas grama			
Texas wintergrass			
Thin paspalum	X		
Toboosagrass			
Tumblegrass			
Tumble windmill grass			
Vine mesquite	X		
Virginia wildrye			
Weeping lovegrass		X	
Western wheatgrass			
White tridens			
Yellow (King Ranch) bluestem			X

TABLE 5.2. Common Names of Exotic Grasses Introduced to Texas That Are Problematic to Northern Bobwhites.

Common name	Origin	Reasons why problematic
Bahiagrass	Mexico, Central and South America	Forms dense stands that inhibit movement and foraging
Bermudagrass	Asia	Forms a dense sod that inhibits movement and foraging, does not provide nesting cover or seeds, chokes out desirable forbs and native grasses
Buffelgrass	India and Africa	Chokes out desirable forbs and native grasses, seeds not eaten, reduces arthropods, often forms dense stands that inhibit movement and foraging
Guineagrass	Tropical Africa, sub-tropics of South Africa	Chokes out desirable forbs and native grasses
Johnsongrass	Mediterranean region and North Africa	Poor nesting cover, chokes out desirable forbs and native grasses
Kleberg's bluestem	Africa and Asia	Chokes out desirable forbs and native grasses, seeds of little value, may inhibit movements and foraging
Kleingrass	Tropical Africa and subtropical South Africa	Kleingrass stands have reduced diversity of desirable native forbs and grasses, Kleingrass seeds are too small to be of much value for food
Lehmann lovegrass	South Africa	Chokes out desirable forbs and native grasses, reduces arthropods, seeds of little value
Tall fescue	Europe	Chokes out desirable forbs and native grasses, seeds of no value, dense stands inhibit movements and foraging
Yellow (King Ranch) bluestem	Asia	Chokes out desirable forbs and native grasses, seeds of little value

Habitat Restoration Guidelines on Sites Bared of Vegetation

Some guidelines to follow when restoring northern bobwhite habitat in areas where the natural vegetation has been destroyed, such as former cropland and pastures, are as follows:

- Plant seeding mixtures that include native bunchgrass that are beneficial for nesting and seed production and avoid planting sod-forming grasses and exotic grasses (table 5.2).
- Include the forbs in this book in planting mixtures if commercially available.
- Plant woody plants that provide good escape and thermal cover in 15–30% of the area. Desirable woody plants include low-growing shrubs such as spiny hackberry and lotebush. Mesquite can be planted with low-growing shrubs to provide additional thermal cover.
- Woody plants are most successfully established by planting seedlings rather than seeds. Seedlings may need to be protected from rodents and deer.
- Woody plants should be planted in groups (mottes) of 10 or more individuals. Mottes should be separated by grasses and forbs and should be no further than the average adult can throw a softball (chapter 4).

In Summary

Grasses are an important component of northern bobwhite habitat, but many exotic grasses are detrimental to bobwhite habitat. Sod-forming grasses such as bermudagrass are particularly detrimental to bobwhites because they impede movement and foraging and do not provide cover for nesting. Also bad for bobwhites are exotic grasses such as buffelgrass and Lehmann lovegrass, which tend to form dense monocultures, exclude forbs and native grasses, and support lower arthropod abundance.

Exotic grasses continue to be planted for livestock forage and erosion control and continue to spread from areas where they are planted. Places that still sustain abundant numbers of wild quail in Texas (chapter 4), particularly the South Texas Plains, will experience population declines in the future with continued planting of exotic forage grasses on soils where they develop dense stands. Steps such as cleaning seeds from vehicles may be essential to help reduce exotic grass invasion in bobwhite habitats where native vegetation continues to dominate.

Continuing degradation of bobwhite habitat from replacement of native-plant communities by exotics is difficult to halt. Methods to replace bermudagrass with native plants are available, but treatments are expensive. Protocols to replace warm-season grasses, including buffelgrass, Lehmann lovegrass, guineagrass, yellow bluestem, and Kleberg's bluestem, with native plants are not currently available. Presently, the only way to improve landscapes dominated by these exotic grasses for bobwhites is to manage vegetation structure through grazing, fire, and mechanical treatments.

Appendix 1

Other Plant Species Seeds Reported in Northern Bobwhite Diets in Texas

SCIENTIFIC NAME	COMMON NAME
Abutilon fruticosum	Texas Indian mallow
Acacia constricta	Whitethorn acacia
Acacia farnesiana	Sweet acacia
Acacia roemeriana	Roundflower catclaw
Acalypha gracilens	Slender threeseed mercury
Acleisanthes obtusa	Berlandier's trumpets
Aeschynomene virginica	Virginia jointvetch
Agalinis spp.	False foxglove
Albizia julibrissin	Silktree
Amaranthus retroflexus	Redroot amaranth
Ambrosia artemisiifolia	Annual ragweed
Amphiachyris dracunculoides	Prairie broomweed
Amphicarpaea bracteata	American hogpeanut
Andropogon spp.	Bluestem
Arachis hypogaea	Peanut
Argythamnia mercurialina	Tall silverbush
Aristida longespica var. *geniculata*	Slimspike threeawn
Arundo donax	Giant reed
Avena sativa	Common oat
Ayenia pilosa	Hairy ayenia
Bernardia myricifolia	Mouse's eye
Bidens spp.	Beggarticks
Bouteloua spp.	Grama
Callirhoe involucrata	Purple poppymallow

Campsis radicans	Trumpet creeper
Carex spp.	Sedge
Celtis occidentalis	Common hackberry
Cerastium glomeratum	Sticky chickweed
Chamaesaracha sordida	Hairy five eyes
Chamaesyce chaetocalyx	Bristlecup sandmat
Chamaesyce fendleri	Fendler's sandmat
Chamaesyce glyptosperma	Ribseed sandmat
Chasmanthium laxum	Slender woodoats
Chenopodium album	Lambsquarters
Chloris × *subdolichostachya*	Shortspike windmill grass
Cissus trifoliata	Sorrelvine
Cocculus diversifolius	Snailseed
Condalia hookeri var. *hookeri*	Brazilian bluewood
Convolvulus arvenis	Field bindweed
Cooperia pedunculata	Prairie lily
Cornus florida	Flowering dogwood
Crataegus viridis	Green hawthorn
Crotalaria spp.	Rattlebox
Croton dioicus	Grassland croton
Croton lindheimerianus	Threeseed croton
Croton michauxii	Michaux's croton
Croton parksii	Parks' croton
Cucurbita foetidissima	Missouri gourd
Cuscuta gronovii	Scaldweed
Cyclachaena xanthifolia	Giant sumpweed
Cylindropuntia leptocaulis	Christmas cactus
Cynanchum barbigerum	Bearded swallow-wort
Cyperus esculentus	Yellow nutsedge
Dalea aurea	Golden prairie clover
Dalea lasiathera	Purple dalea
Daucus spp.	Wild carrot
Desmodium canescens	Hoary ticktrefoil
Desmodium floridanum	Florida ticktrefoil
Desmodium glabellum	Dillenius' ticktrefoil
Desmodium sessilifolium	Sessileleaf ticktrefoil
Dichanthelium oligosanthes	Heller's rosette grass
Dichanthelium sphaerocarpon	Roundseed panicgrass
Digitaria cognata	Fall witchgrass
Digitaria sanguinalis	Hairy crabgrass
Diospyros texana	Texas persimmon

Diospyros virginiana	Common persimmon
Elaeagnus angustifolia	Russian olive
Eleocharis spp.	Spikerush
Eleusine indica	Indian goosegrass
Elymus spp.	Wildrye
Eriochloa spp.	Cupgrass
Euphorbia dentata	Toothed spurge
Euphorbia spathulata	Warty spurge
Evolvulus alsinoides	Slender dwarf morning-glory
Evolvulus sericeus	Silver dwarf morning-glory
Eysenhardtia texana	Texas kidneywood
Forestiera angustifolia	Texas swampprivet
Forestiera pubescens var. *pubescens*	Stretchberry
Fraxinus spp.	Ash
Froelichia drummondii	Drummond's snakecotton
Galactia texana	Texas milkpea
Galactia volubilis	Downy milkpea
Gaura hexandra ssp. *hexandra*	Harlequinbush
Gaura mollis	Velvetweed
Gaura sinuata	Wavyleaf beeblossom
Gossypium spp.	Cotton
Gutierrezia sarothrae	Broom snakeweed
Gutierrezia texana var. *texana*	Texas snakeweed
Helianthus debilis ssp. *cucumerifolius*	Cucumberleaf sunflower
Helianthus praecox	Texas sunflower
Hybanthus verticillatus	Babyslippers
Ibervillea lindheimeri	Lindheimer's globeberry
Ilex decidua	Possumhaw
Ilex glabra	Inkberry
Ilex vomitoria	Yaupon
Ipomoea cordatotriloba var. *cordatotriloba*	Tievine
Iva angustifolia	Narrowleaf marsh elder
Iva annua var. *annua*	Annual marsh elder
Jacquemontia tamnifolia	Hairy clustervine
Jatropha cathartica	Berlandier's nettlespurge
Jatropha dioica	Leatherstem
Juncus spp.	Rush
Kallstroemia californica	California caltrop
Kallstroemia hirsutissima	Hairy caltrop
Krameria ramosissima	Manystem ratany
Krigia caespitosa	Weedy dwarfdandelion

Kummerowia striata	Japanese clover
Lantana canescens	Hammock shrubverbena
Lantana urticoides	West Indian shrubverbena
Lespedeza stuevei	Tall lespedeza
Lespedeza thunbergii	Thunberg's lespedeza
Lespedeza violacea	Violet lespedeza
Lesquerella gracilis	Spreading bladderpod
Lesquerella lasiocarpa	Roughpod bladderpod
Linum virginianum	Woodland flax
Lithospermum incisum	Narrowleaf stoneseed
Lonicera spp.	Honeysuckle
Lotus unifoliolatus var. *unifoliolatus*	American bird's-foot trefoil
Lupinus subcarnosus	Texas bluebonnet
Lycium berlandieri	Berlandier's wolfberry
Maclura pomifera	Osage orange
Magnolia grandiflora	Southern magnolia
Malvastrum coromandelianum	Threelobe false mallow
Medicago polymorpha	Burclover
Melilotus indica	Annual yellow sweetclover
Mimosa malacophylla	Softleaf mimosa
Mimosa microphylla	Littleleaf sensitive-briar
Minuartia drummondii	Drummond's stitchwort
Mollugo verticillata	Green carpetweed
Monarda citriodora	Lemon beebalm
Monarda clinopodioides	Basil beebalm
Monarda punctata	Spotted beebalm
Neptunia lutea	Yellow puff
Neptunia pubescens	Tropical puff
Nyssa sylvatica	Blackgum
Oenothera grandis	Showy evening primrose
Orbexilum pedunculatum var. *psoralioides*	Sampson's snakeroot
Ostrya virginiana	Hophornbeam
Oxalis dichondrifolia	Peonyleaf woodsorrel
Oxalis stricta	Common yellow oxalis
Palafoxia spp.	Palafox
Panicum capillarioides	Slender panicgrass
Panicum hallii	Hall's panicgrass
Panicum obtusum	Vine mesquite
Parkinsonia aculeata	Jerusalem thorn
Parkinsonia texana var. *texana*	Texas paloverde
Parthenocissus quinquefolia	Virginia creeper

Paspalum dilatatum	Dallisgrass
Paspalum floridanum	Florida paspalum
Persea borbonia	Redbay
Phacelia congesta	Caterpillars
Phemeranthus aurantiacus	Orange fameflower
Physalis mollis	Field groundcherry
Pinus palustris	Longleaf pine
Plantago aristata	Largebracted plantain
Plantago hookeriana	California plantain
Plantago lanceolata	Narrowleaf plantain
Polanisia dodecandra ssp. *trachysperma*	Sandyseed clammyweed
Polygonum hydropiperoides	Swamp smartweed
Polygonum punctatum var. *punctatum*	Dotted smartweed
Polygonum scandens var. *dumetorum*	Climbing false buckwheat
Portulaca grandiflora	Rose moss
Portulaca pilosa	Kiss me quick
Prosopis chilensis	Algarrobo
Quercus gambelii	Gambel oak
Quercus nigra	Water oak
Quercus shumardii	Shumard's oak
Quercus sinuata	Bastard oak
Quercus stellata	Post oak
Rhus aromatica	Fragrant sumac
Rhus microphylla	Littleleaf sumac
Rhynchosia minima	Least snoutbean
Rhynchospora spp.	Beaksedge
Richardia brasiliensis	Tropical Mexican clover
Richardia scabra	Rough Mexican clover
Robinia pseudoacacia	Black locust
Rorippa teres	Southern marsh yellowcress
Rosa spp.	Rose
Rubus trivialis	Southern dewberry
Ruellia nudiflora	Violet wild petunia
Ruellia nudiflora var. *runyonii*	Runyon's wild petunia
Rumex acetosella	Common sheep sorrel
Rumex crispus	Curly dock
Rumex hastatulus	Heartwing sorrel
Salvia texana	Texas sage
Sassafras albidum	Sassafras
Schaefferia cuneifolia	Desert yaupon
Scirpus spp.	Bulrush

Scleria oligantha	Littlehead nutrush
Senna bauhinioides	Twinleaf senna
Sesbania herbacea	Bigpod sesbania
Setaria parviflora	Marsh bristlegrass
Setaria reverchonii ssp. *firmula*	Knot grass
Setaria texana	Texas bristlegrass
Sida rhombifolia	Cuban jute
Smilax spp.	Greenbrier
Solanum carolinense	Carolina horsenettle
Solanum nigrum	Black nightshade
Sphaeralcea coccinea	Scarlet globemallow
Sporobolus compositus var. *compositus*	Composite dropseed
Stillingia linearifolia	Queen's-root
Stillingia sylvatica	Queen's-delight
Stillingia treculiana	Trecul's toothleaf
Stylosanthes biflora	Sidebeak pencilflower
Symphoricarpos orbiculatus	Coralberry
Tephrosia lindheimeri	Lindheimer's hoarypea
Tephrosia onobrychoides	Multibloom hoarypea
Tephrosia virginiana	Virginia tephrosia
Thamnosma texana	Rue of the mountains
Toxicodendron vernix	Poison sumac
Tradescantia spp.	Spiderwort
Tragia glanduligera	Brush noseburn
Tragia nepetifolia	Catnip noseburn
Tragia urticifolia	Nettleleaf noseburn
Tridens muticus	Slim tridens
Trifolium carolinianum	Carolina clover
Ulmus crassifolia	Cedar elm
Uniola spp.	Seaoats
Urochloa maxima	Guineagrass
Urochloa platyphylla	Broadleaf signalgrass
Verbena spp.	Vervain
Viburnum spp.	Viburnum
Vicia caroliniana	Carolina vetch
Vicia ludoviciana	Louisiana vetch
Vicia villosa	Winter vetch
Vigna unguiculata	Southern pea
Vitis mustangensis	Mustang grape
Zanthoxylum fagara	Lime pricklyash
Ziziphus obtusifolia	Lotebush

Appendix 2

Other Wildlife That Utilize Seeds Eaten by Northern Bobwhites

Plant	Seeds/Fruit/Mast	Foliage/ Other Plant Parts
RUSHES AND GRASSES		
Scleria ciliata	Songbirds*	
Avena fatua	Rodents Songbirds*	White-tailed Deer
Bromus catharticus	Songbirds*	Rodents White-tailed Deer Aoudad Sheep
Cenchrus spinifex	Rodents Songbirds*	Rodents White-tailed Deer Nilgai*
Dichanthelium spp.	Songbirds* Game Mammals/ Furbearers*	White-tailed Deer*
Echinochloa crus-galli	Waterfowl Songbirds*	Game Mammals/ Furbearers
Panicum capillare	Rodents* White-winged Dove* Waterfowl* Songbirds*	Rodents* White-tailed Deer* Mule Deer* Nilgai*

Panicum virgatum	Rodents*	Rodents*
	White-winged Dove*	White-tailed Deer*
	Waterfowl	Mule Deer*
	Songbirds*	Nilgai*
Paspalum plicatulum	Waterfowl*	Rodents*
	Songbirds*	White-tailed Deer*
		Nilgai*
Paspalum setaceum	Rodents	Rodents*
	Scaled Quail	White-tailed Deer*
	Waterfowl*	Nilgai*
	Prairie Chicken	
	Songbirds*	
Setaria leucopila	Rodents	Javelina*
	Scaled Quail*	Nilgai*
	Gambel's Quail*	White-tailed Deer*
	Mourning Dove*	
	White-winged Dove*	
	Songbirds*	
Sorghum bicolor	Rodents	Game Mammals/
	Scaled Quail	Furbearers
	Mourning Dove	White-tailed Deer
	Wild Turkey	
	Sandhill Crane	
	Whooping Crane	
	Waterfowl	
	Prairie Chicken	
	Songbirds	
	White-tailed Deer	
Sorghum halepense	Rodents	Game Mammals/
	Mourning Dove	Furbearers
	Waterfowl	
	Songbirds	
Triticum aestivum	Scaled Quail	Whooping Crane
	Mourning Dove	Sandhill Crane
	Sandhill Crane	Game Mammals/
	Waterfowl	Furbearers
	Prairie Chicken	White-tailed Deer
	Songbirds	
*Urochloa** ciliatissima*	Rodents*	Rodents*
	White-winged Dove*	White-tailed Deer*
	Waterfowl*	Mule Deer*
	Songbirds*	Nilgai*

*Urochloa** fusca*	Rodents*	Rodents*
	White-winged Dove*	White-tailed Deer*
	Waterfowl*	Mule Deer*
	Songbirds*	Nilgai*
*Urochloa** texana*	Rodents*	Rodents*
	White-winged Dove*	White-tailed Deer*
	Waterfowl*	Mule Deer*
	Songbirds*	Nilgai*
Zea mays	Gambel's Quail	White-tailed Deer
	Mearns' Quail	
	Mourning Dove	
	White-winged Dove	
	Wild Turkey	
	Whooping Crane	
	Sandhill Crane	
	Waterfowl	
	Songbirds	
	Prairie Chicken	
	Game Mammals/	
	Furbearers	
	White-tailed Deer	

FORBS

Amaranthus spp.	Scaled Quail	Game Mammals/
	Mourning Dove	Furbearers
	White-winged Dove	White-tailed Deer
	Wild Turkey	
	Waterfowl	
	Prairie Chicken	
	Songbirds	
Ambrosia psilostachya	Rodents	Rodents
	Scaled Quail	Game Mammals/
	Mourning Dove	Furbearers
	Wild Turkey	Javelina
	Waterfowl	Feral Pig
	Prairie Chicken	White-tailed Deer
	Chachalaca	Pronghorn
	Songbirds	
	Wild Turkey	
Ambrosia trifida	Wild Turkey	
Helianthus annuus	Rodents	Chachalaca
	Scaled Quail*	Game Mammals/

	Gambel's Quail*	Furbearers
	Mearns' Quail*	White-tailed Deer
	Mourning Dove	
	White-winged Dove	
	Wild Turkey	
	Prairie Chicken	
	Songbirds*	
Helianthus argophyllus	Scaled Quail*	White-tailed Deer*
	Gambel's Quail*	
	Mearns' Quail*	
	Mourning Dove*	
	White-winged Dove*	
	Songbirds*	
Heterotheca subaxillaris		Prairie Chicken*
Verbesina encelioides	Gambel's Quail	
	Wild Turkey	
Descurainia pinnata	Gambel's Quail*	Scaled Quail
	Wild Turkey	Javelina
		White-tailed Deer
Commelina erecta	Scaled Quail	Rodents
	Mourning Dove	Prairie Chicken
	White-winged Dove	White-tailed Deer
	Prairie Chicken	Nilgai
Convolvulus equitans	Scaled Quail*	White-tailed Deer
Acalypha radians	Scaled Quail	Rodents
	Mourning Dove*	White-tailed Deer
	Wild Turkey	
	Songbirds*	
Argythamnia humilis	Scaled Quail	White-tailed Deer
	Mourning Dove	Mule Deer
Cnidoscolus texanus	Mourning Dove	
	Wild Turkey	
Croton capitatus	Scaled Quail*	Rodents*
	Mourning Dove	Chachalaca*
	White-winged Dove*	White-tailed Deer*
	Wild Turkey	Mule Deer*
	Prairie Chicken*	Nilgai*
Croton glandulosus	Rodents	Rodents*
	Scaled Quail*	Chachalaca*
	Mourning Dove	White-tailed Deer*
	White-winged Dove*	Mule Deer*

	Prairie Chicken*	Nilgai*
Croton monanthogynus	Scaled Quail Mourning Dove White-winged Dove* Prairie Chicken*	Rodents* Chachalaca* Game Mammals/ Furbearers White-tailed Deer Mule Deer* Nilgai*
Croton punctatus	Scaled Quail* White-winged Dove* Prairie Chicken*	Rodents* Chachalaca* White-tailed Deer* Mule Deer* Nilgai*
Croton texensis	Scaled Quail* Mourning Dove White-winged Dove* Prairie Chicken*	Rodents* Chachalaca* Game Mammals/ Furbearers White-tailed Deer* Mule Deer* Pronghorn Nilgai*
Euphorbia bicolor	Scaled Quail* Gambel's Quail* Mearns' Quail* Songbirds*	Javelina* White-tailed Deer*
Phyllanthus polygonoides	Scaled Quail Mourning Dove White-winged Dove	White-tailed Deer
Astragalus nuttallianus	Scaled Quail Gambel's Quail* Wild Turkey*	Chachalaca* White-tailed Deer*
Chamaecrista fasciculata		White-tailed Deer Nilgai
Chamaecrista flexuosa		White-tailed Deer
Dalea emarginata		White-tailed Deer
Desmanthus illinoensis	Scaled Quail*	
Desmanthus virgatus	Scaled Quail Mourning Dove	White-tailed Deer
Desmodium obtusum	Wild Turkey*	White-tailed Deer*
Galactia canescens	Mourning Dove	White-tailed Deer

	Wild Turkey Songbirds*	
Galactia heterophylla	Songbirds*	White-tailed Deer*
Galactia regularis	Songbirds*	White-tailed Deer*
Indigofera miniata	Mourning Dove	Wild Turkey White-tailed Deer
Kummerowia stipulacea	Wild Turkey*	Game Mammals/ Furbearers* White-tailed Deer*
Lespedeza virginica	Wild Turkey	White-tailed Deer
Lupinus texensis	Scaled Quail* Gambel's Quail*	Rodents
Mimosa latidens	Mourning Dove	White-tailed Deer
Rhynchosia americana		White-tailed Deer Nilgai*
Rhynchosia latifolia		Nilgai*
Strophostyles helvola	Mourning Dove	White-tailed Deer
Strophostyles leiosperma	Mourning Dove	White-tailed Deer
Strophostyles umbellata	Mourning Dove	White-tailed Deer
Vicia ludoviciana ssp. *leavenworthii*	Scaled Quail* Songbirds*	White-tailed Deer Aoudad Sheep
Zornia bracteata		White-tailed Deer
Geranium carolinianum	Mourning Dove Waterfowl* Songbirds	Wild Turkey White-tailed Deer
Mentzelia nuda	Gambel's Quail	White-tailed Deer* Pronghorn
Sida abutifolia	Scaled Quail	White-tailed Deer
Menodora heterophylla	Scaled Quail	White-tailed Deer Mule Deer*
Oxalis dillennii	Mearns' Quail* Waterfowl* Songbirds* Chachalaca	Wild Turkey Javelina White-tailed Deer
Argemone sanguinea	Mourning Dove	White-tailed Deer*
Plantago rhodosperma	Scaled Quail	Rodents*

	Mourning Dove	Javelina*
	Game Mammals/ Furbearers	White-tailed Deer
		Aoudad Sheep
		Texas Tortoise
Diodia teres	Scaled Quail	White-tailed Deer
	Wild Turkey	
	Waterfowl	
	Prairie Chicken	
Physalis cinerascens	Mourning Dove	Javelina
	Wild Turkey	White-tailed Deer
	Chachalaca*	
	Game Mammals/ Furbearers*	
Solanum elaeagnifolium	Rodents	Game Mammals/ Furbearers
	Scaled Quail*	
	White-winged Dove*	Wild Turkey
	Songbirds*	Javelina*
	Javelina	White-tailed Deer
	Feral Pig	Nilgai*
	White-tailed Deer	
Solanum rostratum	Rodents	Javelina*
	Scaled Quail*	Nilgai*
	White-winged Dove*	
	Wild Turkey	
	Songbirds*	

WOODY VINES

Toxicodendron radicans	Wild Turkey	White-tailed Deer
	Songbirds	
Ampelopsis arborea	Game Mammals/ Furbearers	White-tailed Deer

WOODY PLANTS

Mahonia trifoliolata		White-tailed Deer
Tiquilia canescens	Scaled Quail	
Acacia rigidula	White-winged Dove*	White-tailed Deer
	Chachalaca	Mule Deer*
		Javelina
Prosopis glandulosa	Rodents	Rodents
	Scaled Quail	Chachalaca

	Gambel's Quail*	Javelina
	Wild Turkey	White-tailed Deer
	Chachalaca	Mule Deer
	Game Mammals/	Pronghorn
	Furbearers	Nilgai
	Javelina	
	White-tailed Deer	
Quercus virginiana	Mearns' Quail*	Nilgai
	Wild Turkey*	Mule Deer*
	Sandhill Crane	White-tailed Deer
	Waterfowl	Aoudad Sheep
	Songbirds*	
	Game Mammals/	
	Furbearers	
	Javelina*	
	Feral Pig	
	White-tailed Deer	
Liquidambar styraciflua	Waterfowl	
	Game Mammals/	
	Furbearers	
Morella cerifera	Waterfowl	White-tailed Deer
	Songbirds*	
	Game Mammals/	
	Furbearers*	
Pinus spp.	Mourning Dove	
	Songbirds	
Sideroxylon lanuginosum	White-winged Dove*	White-tailed Deer
	Wild Turkey	
	Chachalaca	
	Game Mammals/	
	Furbearers	
Celtis ehrenbergia	Rodents*	Game Mammals/
	Scaled Quail	Furbearers*
	Wild Turkey*	Javelina
	Chachalaca	White-tailed Deer
	Songbirds*	Mule Deer
	Game Mammals/	Aoudad Sheep*
	Furbearers*	Nilgai
Celtis laevigata	Rodents*	Game Mammals/
var. *reticulata*	Scaled Quail*	Furbearers*
	Wild Turkey*	Chachalaca

	Chachalaca	White-tailed Deer
	Songbirds*	Aoudad Sheep*
	Game Mammals/	
	Furbearers*	
Callicarpa americana	Rodents	White-tailed Deer
	Songbirds	
	Game Mammals/	
	Furbearers	
	White-tailed Deer	

SUCCULENTS

Opuntia engelmannii	Rodents	Javelina
var. *lindheimeri*	Scaled Quail*	White-tailed Deer
	Gambel's Quail*	Mule Deer
	Mearns' Quail*	Pronghorn*
	White-winged Dove*	
	Wild Turkey	
	Prairie Chicken*	
	Songbirds*	
	Game Mammals/	
	Furbearers	
	Javelina	
	White-tailed Deer	

No references on use by wildlife other than bobwhites were found for *Centrosema virginianum, Chamaecrista calycioides, Chamaecrista nictitans, Justicia pilosella, Pediomelum rhombifolium,* and *Zanthoxylum clava-herculis.*

Game Mammals/Furbearers include rabbits, raccoons, coyotes, Ring-tailed Cats, opossums, foxes, armadillos, beavers, skunks, and squirrels.

* Indicates use by a wildlife species has been recorded for the plant genus, but not the plant species.

** Plants in this genus were changed from the genus *Panicum,* so wildlife use for this genus is generally the same as *Panicum* spp.

Appendix 3

Common and Scientific Names of Plants and Animals Mentioned in Text

Common Name	Scientific Name
PLANTS	
Alkali sacaton	*Sporobolus airoides*
Arizona cottontop	*Digitaria californica*
Bahiagrass	*Paspalum notatum*
Beaked panicgrass	*Panicum anceps*
Bermudagrass	*Cynodon dactylon*
Big bluestem	*Andropogon gerardii*
Big sandbur	*Cenchrus myosuroides*
Black grama	*Bouteloua eriopoda*
Blue grama	*Bouteloua gracilis*
Blue panicum	*Panicum antidotale*
Broomsedge bluestem	*Andropogon virginicus*
Buffalograss	*Bouteloua dactyloides*
Buffelgrass	*Pennisetum ciliare*
Burrograss	*Scleropogon brevifolius*
Bush muhly	*Muhlenbergia porteri*
Canada wildrye	*Elymus canadensis*
Cane bluestem	*Bothriochloa barbinodis*
Caucasian bluestem	*Bothriochloa bladhii*
Common carpetgrass	*Axonopus fissifolius*
Composite dropseed	*Sporobolus compositus* var. *macer*
Cotton	*Gossypium* spp.
Curly-mesquite	*Hilaria belangeri*
Cylinder jointtail grass	*Coelorachis cylindrical*
Dallisgrass	*Paspalum dilatatum*

Common name	Scientific name
Eastern gamagrass	*Tripsacum dactyloides*
Fall witchgrass	*Digitaria cognata*
Florida paspalum	*Paspalum floridanum*
Green sprangletop	*Leptochloa dubia*
Guajillo	*Acacia berlandieri*
Guineagrass	*Urochloa maxima*
Gulf cordgrass	*Spartina spartinae*
Hairy grama	*Bouteloua hirsuta*
Hairy woollygrass	*Erioneuron pilosum*
Hall's panicgrass	*Panicum hallii* var. *hallii*
Heller's rosette grass	*Dichanthelium oligosanthes*
Hooded windmill grass	*Chloris cucullata*
Indiangrass	*Sorghastrum nutans*
Kleberg's bluestem	*Dichanthium annulatum*
Kleingrass	*Panicum coloratum*
Lehmann lovegrass	*Eragrostis lehmanniana*
Little barley	*Hordeum pusillum*
Little bluestem	*Schizachyrium scoparium*
Longtom	*Paspalum denticulatum*
Lotebush	*Ziziphus obtusifolia*
Marsh bristlegrass	*Setaria parviflora*
Multiflower false Rhodes grass	*Trichloris pluriflora*
Pearl millet	*Pennisetum glaucum*
Perennial ryegrass	*Lolium perenne*
Pink pappusgrass	*Pappophorum bicolor*
Plains lovegrass	*Eragrostis intermedia*
Prairie threeawn	*Aristida oligantha*
Purple lovegrass	*Eragrostis spectabilis*
Purple threeawn	*Aristida purpurea*
Purpletop tridens	*Tridens flavus*
Red grama	*Bouteloua trifida*
Red lovegrass	*Eragrostis secundiflora*
Rhodes grass	*Chloris gayana*
Rose Natal grass	*Melinis repens*
Saltgrass	*Distichlis spicata*
Saltmeadow cordgrass	*Spartina patens*
Sand dropseed	*Sporobolus cryptandrus*
Sand lovegrass	*Eragrostis trichodes*
Sideoats grama	*Bouteloua curtipendula*
Silver beardgrass	*Bothriochloa laguroides*
Smut grass	*Sporobolus indicus*
Soybean	*Glycine max*
Splitbeard bluestem	*Andropogon ternarius*
Tall fescue	*Schedonorus phoenix*
Tanglehead	*Heteropogon contortus*
Texas bluegrass	*Poa arachnifera*

Texas cupgrass	*Eriochloa sericea*
Texas grama	*Bouteloua rigidiseta*
Texas wintergrass	*Nassella leucotricha*
Tobosagrass	*Pleuraphis mutica*
Tumblegrass	*Schedonnardus paniculatus*
Tumble windmill grass	*Chloris verticillata*
Vine mesquite	*Panicum obtusum*
Virginia widrye	*Elymus virginicus*
Weeping lovegrass	*Eragrostis curvula*
Western wheatgrass	*Pascopyrum smithii*
White tridens	*Tridens albescens*
Wilman lovegrass	*Eragrostis superba*
Yellow (King Ranch) bluestem	*Bothriochloa ischaemum* var. *songarica*

ANIMALS

Aoudad Sheep	*Ammotragus lervia*
Armadillo	*Dasypus novemcinctus*
Beaver	*Castor canadensis*
Chachalaca	*Ortalis vetula*
Coyote	*Canis latrans*
Feral Pig	*Sus scrofa*
Gambel's Quail	*Callipepla gambelii*
Gray Fox	*Urocyon cinereoargenteus*
Javelina	*Pecari tajacu*
Mearns' (Montezuma) Quail	*Cytonyx montezuma*
Mourning Dove	*Zenaida macroura*
Mule Deer	*Odocoileus hemionus*
Nilgai	*Boselaphus tragocamelus*
Opossum	*Didelphis virginiana*
Prairie Chicken	*Tympanuchus pallidicinctus*
Pronghorn	*Antilocapra americana*
Rabbit	*Sylvilagus* spp.
Raccoon	*Procyon lotor*
Red-cockaded Woodpecker	*Picoides borealis*
Red Fox	*Vulpes vulpes*
Ring-tailed Cat	*Bassariscus astutus*
Sandhill Crane	*Grus canadensis*
Scaled Quail	*Callipepla squamata*
Skunk	*Mepitidae*
Squirrel	*Sciurus* spp.
White-tailed Deer	*Odocoileus virginianus*
White-winged Dove	*Zenaida asiatica*
Whooping Crane	*Grus americana*
Wild Turkey	*Meleagris gallopavo*

Glossary

Achene A small, dry, indehiscent fruit containing one seed.

Acorn Hard, dry, indehiscent fruit of oaks containing a single seed usually seated in a woody cuplike structure.

Alternate Referring to leaves not opposite to each other, but originating singly at different points along a stem.

Angled Referring to a plant part that has evident ridges or sides.

Annual A plant that completes its life cycle in one growing season or year.

Arthropod Invertebrates in the phylum Arthropoda; includes spiders, insects, and crabs.

Awn A bristlelike appendage found on the spikelets of some grasses.

Berry A fleshy fruit with several seeds.

Biennial A plant that completes its life cycle in two growing seasons or years.

Bract A modified leaf or leaflike part just below a flower or inflorescence.

Branching An outgrowth or major division of a stem or trunk.

Bristle A short, stiff hairlike appendage.

Bunchgrass Grasses that grow in the form of a clump rather than spreading as a sod.

Calcareous Containing calcium or limestone, as in soils.

Caliche A hard deposit of calcium carbonate on rocky soils in arid regions.

Canopy cover Amount of ground surface beneath vegetation foliage.

Capsule A dry, dehiscent fruit composed of more than one carpel.

Carpel The unit that makes up the female reproductive organ of a flower.

Caruncle An extension or protuberance at or near the hilum of a seed.

Clay Soil composed of very fine minerals.

Compound Referring to a leaf with one or more leaflets attached to a single leaf stem.

Compressed Flattened, usually laterally or lengthwise.

Concave Curving inward, as the inside of a sphere.

Cone A reproductive structure of plants such as conifers, typically consisting of overlapping scales, which bear seeds.

Conservation Reserve Program A program sponsored by the United States Department of Agriculture to establish perennial plant cover on lands subject to erosion.

Corky Referring to a nonliving, soft tissue formed on the bark of some plants.

Creeping Growing along the ground, often rooting at nodes.

Cylindrical Resembling a cylinder.

Deciduous Referring to the falling or loss of leaves annually in some plants.

Dehiscent Breaking open at maturity to release the contents.

Discing Breaking the soil surface with a plow composed of circular blades.

Dissemination Spreading or scattering of seeds.

Drooping Bending downward.

Drupe A fleshy fruit, usually containing a single seed.

Ecosystem Includes living organisms and their non-living surroundings and the interactions between them.

Elliptic In the shape of a narrow oval.

Erect Straight and upright in position.

Evergreen Having green leaves throughout the year.

Exotic A plant not historically originating or growing naturally in a specific region.

Follicle A single-chambered fruit that splits along one seam to release its seeds.

Forb An herbaceous or non-woody plant other than a grass.

Furrowed Having longitudinal ridges and grooves.

Globose In the shape of a sphere.

Granivorous An organism that consumes a diet composed of seeds.

Gravel Soil type composed mostly of small rocks or gravel.

Hairy Covered with fine to stiff hairs.

Hilum The scar on a seed from its point of attachment.

Inconspicuous Difficult to see.

Indehiscent Not opening at maturity.

Inflorescence The flowering part of a plant or the arrangement of flowers on a stalk.

Introduced A plant that does not historically originate or grow in a specific region.

Invasive A plant prone to aggressively spread from areas where it has been planted into areas where it was not planted.

Leaflet A division of a compound leaf.

Legume A dry, dehiscent fruit usually opening up along two sides releasing seeds. Also referring to a plant belonging to the family Fabaceae.

Lemma The lower of two bracts surrounding the flower in grasses.

Lignify To become woody.

Linear Much longer than broad, with sides mostly parallel, as in a leaf.

Loam Soil composed of relatively equal concentrations of clay, sand, and silt.

Lobe A usually rounded segment or projection of a plant part, as in a leaf.

Longitudinal Along the long axis of a plant part.

Margin The edge of a flat structure, such as a leaf.

Mast Fruits of trees and shrubs.

Monoculture A stand of plants composed of only one species.

Motte Referring to a group or cluster of trees, typically live oaks.

Mottled Colored with spots or blotches of different colors.

Native A plant historically originating or growing naturally in a specific region.

Node A point on a stem where a leaf or branch originates.

Nutlet A small nut.

Oblong Elongated.

Opposite Referring to a pair of leaves that originate at the same axis on a stem, but opposite each other.

Ovate Egg-shaped.

Palea The upper of two bracts surrounding the flower in grasses.

Perennial A plant that lives for three or more years.

Petal A single segment of the corolla (all petals of a flower).

Plant community A group of plants that occur together under a particular set of environmental conditions.

Prickle A small, sharp spine.

Protuberance Something that bulges out.

Rangeland Uncultivated land that supports the needs of browsing and grazing animals. Includes grasslands, shrublands, deserts, savannahs, and open forests.

Reclining Curving or bending downward.

Rhizomatous A plant with subterranean stems that spread laterally beneath the ground and are capable of producing shoots and adventitious roots at the nodes.

Ribbed With prominent longitudinal ridges, as on a stem.

Ruderal Vegetation growing on disturbed soils and showing a preference for this type of habitat.

Sand Soil composed mostly of sand particles.

Sedge A grasslike member of the Cyperaceae family.

Shrub A woody plant, with several stems, of relatively low height.

Silky Covered with fine soft hairs.

Silt Soil composed mostly of very fine sedimentary particles intermediate in size between clay and sand.

Simple Referring to a leaf that is undivided or not separated into leaflets.

Spherical Resembling a sphere.

Spikelet A small or secondary spike, characteristic of grasses and sedges, having a varying number of reduced flowers, each subtended by one or two scalelike bracts.

Spine A rigid, slender, sharp projection originating from a plant part.

Spreading A plant or plant part that extends horizontally more so than vertically.

Subshrub A perennial plant, similar to a shrub, but typically woody at only the lower portions.

Subtend To enclose or surround.

Subterranean Below the ground.

Succulent A plant having fleshy parts, as in a cactus, usually adapted to arid environments.

Terminal At the tip.

Thicket A dense growth of shrubs.

Toothed A leaf margin divided into small lobes or tooth-like segments.

Trailing Creeping or spreading but not rooting.

Tree A large woody plant, typically with a single stem or trunk.

Twig A small shoot or branch from a tree or shrub.

Twining Encircling or coiling around a support, typically another stem, as in a vine.

Utricle A small, thin-walled, slightly inflated one-seeded fruit.

Vascular bundle A channel of tissue in plants that conveys fluids.

Vein A vascular bundle, usually visible, as in a leaf.

Vine A plant without a self-supporting stem, usually climbing or trailing on some support.

Warty Covered with hard protuberances.

Waxy Covered with a smooth wax coating.

Wilt To become limp or soft.

Woody A plant with hardened, lignified outer tissue (bark) on stems or trunks.

Wooly Covered with soft, fine, entangled hairs.

References

Ajilvsgi, G. 1979. *Wild Flowers of the Big Thicket, East Texas and Western Louisiana.* College Station: Texas A&M University Press. 360 pp.

———. 1984. *Wildflowers of Texas.* Fredericksburg, Tex.: Shearer Publishing. 524 pp.

Alcoze, T. M., and E. G. Zimmerman. 1973. Food habits and dietary overlap of two Heteromyid rodents from the mesquite plains of Texas. Journal of Mammalogy 54:900–908.

Allaire, P. N., and C. D. Fisher. 1975. Feeding ecology of three resident sympatric sparrows in eastern Texas. The Auk 92:260–269.

Allen, C. E. 1980. Feeding habits of ducks in a green-tree reservoir in eastern Texas. Journal of Wildlife Management 44:232–236.

Angell, D. L., and M. P. McClaran. 2001. Long-term influences of livestock management and a non-native grass on grass dynamics in the Desert Grassland. Journal of Arid Environments 49:507–520.

Archer, S. R., and K. I. Predick. 2008. Climate change and ecosystems of the southwestern United States. Rangelands 30:23–28.

Arredondo, J. A., F. Hernández, F. C. Bryant, R. L. Bingham, and R. Howard. 2007. Habitat-suitability bounds for nesting cover of northern bobwhites on semiarid rangelands. Journal of Wildlife Management 71:2592–2599.

Arriaga, L., A. E. Castellanos, E. Moreno, and J. Alarcón. 2004. Potential ecological distribution of alien invasive species and risk assessment: a case study of buffel grass in arid regions of Mexico. Conservation Biology 18:1504–1514.

Ault, S. C., and F. A. Stormer. 1983. Seasonal food selection by scaled quail in northwest Texas. Journal of Wildlife Management 47:222–228.

Baker, R. H., C. C. Newman, and F. Wilke. 1945. Food habits of the raccoon in eastern Texas. Journal of Wildlife Management 9:45–48.

Ballard, B. M., and J. E. Thompson. 2000. Winter diets of sandhill cranes from central and coastal Texas. Wilson Bulletin 112:263–268.

Barclay, A. S., and F. R. Earle. 1974. Chemical analyses of seeds III: oil and protein content of 1253 species. Economic Botany 28:176–236.

Barnes, T. G. 2004. Strategies to convert exotic grass pastures to tall grass prairie communities. Weed Technology 18:1364–1370.

Boren, J. C., R. L. Lochmiller, D. M. Leslie, Jr., and D. M. Engle. 1995. Amino acid concentrations in seed of preferred forages of bobwhites. Journal of Range Management 48:141–144.

Brennan, L. A., editor. 2007. *Texas Quails: Ecology and Management*. College Station: Texas A&M University Press. 491 pp.

Burton, G. W., and P. R. Utley. Not dated. Forage and turfgrass research. Bermudagrass for forage. http://www.tifton.uga.edu/fat/bermudagrass.htm, accessed 18 March 2008.

Campbell-Kissock, L., L. H. Blankenship, and J. W. Stewart. 1985. Plant and animal foods of bobwhite and scaled quail in southwest Texas. Southwestern Naturalist 30:543–553.

Case, R. M., and R. J. Robel. 1974. Bioenergetics of the bobwhite. Journal of Wildlife Management 38:638–652.

Chamrad, A. D., and T. W. Box. 1968. Food habits of white-tailed deer in south Texas. Journal of Range Management 21:158–164.

Chapman, B. R. 1972. Food habits of Loring's kangaroo rat, *Dipodomys elator*. Journal of Mammalogy 53:877–880.

Chenault, T. P. 1940. The phenology of some bob-white food and cover plants in Brazos County, Texas. Journal of Wildlife Management 4:359–368.

Christensen, Z. D., D. B. Pence, and G. Scott. 1978. Notes on food habits of the plain chachalaca from the lower Rio Grande Valley. Wilson Bulletin 90:647–648.

Corn, J. L., and R. J. Warren. 1985. Seasonal food habits of the collared peccary in south Texas. Journal of Mammalogy 66:155–159.

Correl, D. S., and M. C. Johnston. 1970. *Manual of the Vascular Plants of Texas*. Renner, Tex.: Texas Research Foundation. 1881 pp.

Cox, J. R., M. H. Martin-R, F. A. Ibarra-F, J. H. Fourie, N. F. G. Rethman, and D. G. Wilcox. 1988. The influence of climate and soils on the distribution of four African grasses. Journal of Range Management 41:127–139.

Crawford, J. A., and E. G. Bolen. 1976. Fall diet of lesser prairie chickens in west Texas. Condor 78:142–144.

Cutter, W. L. 1958. Food habits of the swift fox in northern Texas. Journal of Mammalogy 39:527–532.

Davis, C. A., and L. M. Smith. 1998. Ecology and management of migrant shorebirds in the Playa Lakes region of Texas. Wildlife Monographs 140:3–45.

Davis, L. W. 1993. *Weed Seeds of the Great Plains*. Lawrence: University Press of Kansas. 145 pp.

Dietz, D. R. 1999. Winter food habits and preferences of northern bobwhite in east Texas. M.S. Thesis. Nacogdoches, Tex.: Stephen F. Austin State University. 105 pp.

Doerr, T. B., and F. S. Guthery. 1983. Food selection by lesser prairie chickens in northwest Texas. Southwestern Naturalist 28:381–383.

Drawe, D. L. 1968. Mid-summer diet of deer on the Welder Wildlife Refuge. Journal of Range Management 21:164–166.

Earle, F. R., and Q. Jones. 1962. Analyses of seed samples from 113 plant families. Economic Botany 16:221–250.

Enquist, M. 1987. *Wildflowers of the Texas Hill Country*. Austin: Lone Star Botanical. 275 pp.

Everitt, J. H., and D. L. Drawe. 1974. Spring food habits of white-tailed deer in the south Texas plains. Journal of Range Management 27:15–20.

Everitt, J. H., D. L. Drawe, and R. I. Lonard. 1999. *Field Guide to the Broad-Leaved Herbaceous Plants of South Texas Used by Livestock and Wildlife*. Lubbock: Texas Tech University Press. 277 pp.

———. 2002. *Trees, Shrubs and Cacti of South Texas*. Lubbock: Texas Tech University Press. 249 pp.

Everitt, J. H., and M. A. Alaniz. 1980. Fall and winter diets of feral pigs in south Texas. Journal of Range Management 33:126–129.

Everitt, J. H., R. I. Lonard, and C. R. Little. 2007. *Weeds in South Texas and Northern Mexico*. Lubbock: Texas Tech University Press. 240 pp.

Flanders, A. A., W. P. Kuvlesky, Jr., D.C. Ruthven III, R. E. Zaiglan, R. L. Bingham, T. E. Fulbright, F. Hernández, and L. A. Brennan. 2006. Impacts of exotic grasses on south Texas rangeland breeding birds. The Auk 123:171–182.

Gabbard. B. L., and N. L. Fowler. 2007. Wide ecological amplitude of a diversity-reducing exotic grass. Biological Invasions 9:149–160.

Germano, D. J., G. B. Rathbun, and L. R. Saslaw. 2001. Managing exotic grasses and conserving declining species. Wildlife Society Bulletin 29:551–559.

Giuliano, W. M., R. S. Lutz, and R. Patiño. 1996. Reproductive responses of adult female northern bobwhite and scaled quail to nutritional stress. Journal of Wildlife Management 60:302–309.

Glazener, W. C. 1946. Food habits of wild geese on the gulf coast of Texas. Journal of Wildlife Management 10:322–329.

Gonzalez, C. L., and J. D. Dodd. 1979. Production response of native and introduced grasses to mechanical brush manipulation, seeding, and fertilization. Journal of Range Management 32:305–309

Gould, F. W. 1975. *The Grasses of Texas*. College Station: Texas A&M University Press. 653 pp.

———. 1978. *Common Texas Grasses: An Illustrated Guide*. College Station: Texas A&M University Press. 267 pp.

Grace, J., M. Smith, S. Grace, S. Collins, and T. Stohlgren. 2000. Interactions between fire and invasive plants in temperate grasslands of North America. San Diego: Invasive Species Workshop: The Role of Fire in the Control and Spread of Invasive Species. Special Session of the Fire Conference 2000: The First National Congress on Fire Ecology, Prevention and Management.

Gruchy, J. P. 2007. An evaluation of field management practices to improve bobwhite habitat. M.S. Thesis, Knoxville: University of Tennessee. 138 pp.

Guthery, F. S. 1975. Food habits of sandhill cranes in southern Texas. Journal of Wildlife Management 39:221–223.

———. 1986. *Beef Brush and Bobwhites*. Kingsville, Tex.: CKWRI Press. 182 pp.

———. 2000. *On Bobwhites*. College Station: Texas A&M University Press. 213 pp.

———. 2002. *Technology of Bobwhite Management: The Theory Behind the Practice*. Ames: Iowa State University Press. 224 pp.

Hanselka, C. W., and F. S. Guthery. 1991. Bobwhite quail management in south Texas. College Station: Texas Agricultural Extension Service, Texas A&M University System, Publication B-5005. 8 pp.

Harmoney, K. R., P. W. Stahlman, and K. R. Hickman. 2007. Suppression of caucasian old world bluestem with split application of herbicides. Weed Technology 21:573–577.

Harper, C. A. 2007. Strategies for managing early succession habitat for wildlife. Weed Technology 21:932–937.

Hatch, S. L., and J. Pluhar. 1993. *Texas Range Plants*. College Station: Texas A&M University Press. 326 pp.

Hatch, S. L., J. L. Schuster, and D. L. Drawe. 1999. *Grasses of the Texas Gulf Prairies and Marshes*. College Station: Texas A&M University Press. 355 pp.

Hays, K. B., M. Wagner, F. Smeins, and R. N. Wilkins. 2005. Restoring native grasslands. College Station: Texas Cooperative Extension, The Texas A&M University System. 4 pp.

Hellickson, M., and A. Radomski. 1999. Bobwhites of the Wild Horse Desert: status of our knowledge. Caesar Kleberg Wildlife Research Institute Management Bulletin Number 4. 11 pp.

Hickman, K. R., G. H. Farley, R. Channell, and J. E. Steier. 2006. Effects of old world bluestem (*Bothriochloa ischaemum*) on food availability and avian community composition within the mixed-grass prairie. Southwestern Naturalist 51:524–530.

Holm, L., J. V. Pancho, J. P. Herberger, and D. L. Plucknett. 1979. *A Geographical Atlas of World Weeds*. New York: John Wiley. 391 pp.

Ibarra-F, F. A., M. H. Martin-R, T. A. Crowl, and C. A. Call. 1995. Predicting buffelgrass survival across a geographical and environmental gradient. Journal of Range Management 48:53–59.

Jackson, A. S. 1969. A handbook for bobwhite quail management in the west Texas rolling plains. Bulletin 48. Austin: Texas Parks and Wildlife Department. 77 pp.

Jackson, A. S., and R. DeArment. 1963. The lesser prairie chicken in the Texas panhandle. Journal of Wildlife Management 27:733–737.

Jackson, A. S., C. Holt, and D. W. Lay. 1987. Bobwhite quail in Texas: habitat needs and management suggestions. Brochure 7100-37. Austin: Texas Parks and Wildlife Department.

Johnson, M. V. V., and T. E. Fulbright. 2008. Is exotic plant invasion enhanced by a traditional wildlife habitat management technique? Journal of Arid Environments 72:1911–1917.

Jones, F. B., C. M. Rowell, Jr., and M. C. Johnston. 1961. *Flowering Plants and Ferns of the Texas Coastal Bend Counties*. Sinton, Tex.: Rob and Bessie Welder Wildlife Foundation. 146 pp.

Jones, Q., and F. R. Earle. 1966. Chemical analyses of seeds II: oil and protein content of 759 species. Economic Botany 20:127–155.

Kie, J. G., D. L. Drawe, and G. Scott. 1980. Changes in diet and nutrition with increased herd size in Texas white-tailed deer. Journal of Range Management 33:28–34.

Kirkpatrick, Z. M. 1992. *Wildflowers of the Western Plains*. Austin: University of Texas Press. 240 pp.

Koerth, B. H., L. J. Krysl, B. F. Sowell, and F. C. Bryant. 1984. Estimating seasonal diet quality of pronghorn antelope from fecal analysis. Journal of Range Management 37:560–564.

Koerth, B. H., J. L. Mutz, and J. C. Segers. 1986. Availability of bobwhite foods after burning of Pan-American balsamscale. Wildlife Society Bulletin 14:146–150.

Krausman, P. R. 1978. Forage relationships between two deer species in Big Bend National Park, Texas. Journal of Wildlife Management 42:101–107.

Kuvlesky, W. P., Jr., T. E. Fulbright, and R. Engel-Wilson. 2002. The impact of invasive exotic grasses on quail in the southeastern United States. In S. J. DeMaso, W. P. Kuvlesky, Jr., F. Hernández, and M. E. Berger, eds. Quail V: The 5th National Quail Symposium. Pp. 118–128. Austin: Texas Parks and Wildlife Department.

Lay, D. W. 1940. Bobwhite populations as affected by woodland management in eastern Texas. Bulletin No. 592. College Station: Texas Agricultural Experimental Station.

———. 1954. Quail management handbook for east Texas. Bulletin No. 34. Austin: Texas Parks and Wildlife Department. 47 pp.

Leif, A. P., and L. M. Smith. 1993. Winter diet quality, gut morphology and condition of northern bobwhite and scaled quail in west Texas. Journal of Field Ornithology 64:527–538.

Lehman, R. L., R. O'Brien, and T. White. 2005. *Plants of the Texas Coastal Bend*. College Station: Texas A&M University Press. 352 pp.

Lehmann, V. W. 1984. *Bobwhites in the Rio Grande Plain of Texas.* College Station: Texas A&M University Press. 371 pp.

Lehmann, V. W., and H. Ward. 1941. Some plants valuable to quail in southwestern Texas. Journal of Wildlife Management 5:131–135.

Leithead, H. L., L. L. Yarlett, and T. N. Shiflet. 1971. *100 Native Forage Grasses in 11 Southern States.* Agricultural Handbook No. 389, U.S.D.A. Washington, D.C.: Soil Conservation Service. 216 pp.

Loflin, B., and S. Loflin. 2006. *Grasses of the Texas Hill Country.* College Station: Texas A&M University Press. 195 pp.

Lusk, J. L., S. G. Smith, S. D. Fuhlendorf, and F. S. Guthery. 2006. Factors influencing northern bobwhite nest-site selection and fate. Journal of Wildlife Management 70:564–571.

Mahler, W. F. 1988. *Shinners' Manual of the North Central Texas Flora.* Fort Worth: Botanical Research Institute of Texas, Inc. 313 pp.

Marion, W. R. 1976. Plain chachalaca food habits in south Texas. The Auk 93:376–379.

Martin, A. C., H. S. Zim, and A. L. Nelson. 1951. *American Wildlife & Plants: A Guide to Wildlife Food Habits.* New York: Dover Publications. 500 pp.

McAtee, W. L. 1922. Notes on food habits of the shoveller or spoonbill duck (*Spatula clypeata*). The Auk 39:380–386.

McGlone, C. M., and L. F. Huenneke. 2004. The impact of a prescribed burn on Lehmann lovegrass versus native vegetation in the northern Chihuahuan desert. Journal of Arid Environments 57:297–310.

McIntyre, N. E., and T. R. Thompson. 2003. A comparison of Conservation Reserve Program habitat plantings with respect to arthropod prey for grassland birds. American Midland Naturalist 150:291–301.

McMahan, C. A. 1964. Comparative food habits of deer and three classes of livestock. Journal of Wildlife Management 28:798–808.

Meinzer, W. P., D. N. Ueckert, and J. T. Flinders. 1975. Foodniche of coyotes in the rolling plains of Texas. Journal of Range Management 28:22–27.

Miller, J. H., and K. V. Miller. 1999. *Forest Plants of the Southeast and Their Wildlife Uses.* Auburn, Alabama: Southern Weed Science Society. 454 pp.

Nestler, R. B. 1945. Value of wild feedstuffs for pen-reared bobwhite quail in winter. Journal of Wildlife Management 9:115–120.

———. 1949. Nutrition of bobwhite quail. Journal of Wildlife Management 13:342–358.

Nestler, R. B., and W. W. Bailey. 1943. Vitamin A deficiency in bobwhite quail. Journal of Wildlife Management 7:170–173.

Padley, F. B., F. D. Gunstone, and J. L. Harwood. 1994. *The Lipid Handbook.* London: Chapman and Hall/CRC. 1273 pp.

Parmalee, P. W. 1952. Contribution to the ecology of bobwhite quail in the post-oak region of Texas. Unpublished Ph.D. Dissertation. College Station: Texas A&M University. 166 pp.

————. 1953. Food and cover relationships of the bobwhite quail in east-central Texas. Ecology 34:758–770.

————. 1955. Notes on the winter foods of bobwhites in north-central Texas. Texas Journal of Science 7:189–195.

Peoples, A. D., R. L. Lochmiller, J. C. Boren, D. M. Leslie, Jr., and D. M. Engle. 1994. Limitations of amino acids in diets of northern bobwhites (*Colinus virginianus*). American Midland Naturalist 132:104–116.

Peoples, A. D., R. L. Lochmiller, D. M. Leslie, Jr., J. C. Boren, and D. M. Engle. 1994. Essential amino acids in northern bobwhite foods. Journal of Wildlife Management 58:167–175.

Powell, A. M. 1998. *Trees and Shrubs of the Trans-Pecos and Adjacent Areas.* Austin: University of Texas Press. 498 pp.

Quinton, D. A., R. G. Horejsi, and J. T. Flinders. 1979. Influence of brush control on white-tailed deer diets in north-central Texas. Journal of Range Management 32:93–97.

Quinton, D. A., and A. K. Montei. 1977. Preliminary study of the diet of Rio Grande turkeys in north-central Texas. Southwestern Naturalist 22:550–553.

Ramsey, C. W., and M. T. Anderegg. 1972. Food habits of an aoudad sheep, *Ammotragus lervia* (Bovidae), in the Edwards Plateau of Texas. Southwestern Naturalist 16:267–280.

Reeves, R. G. 1972. *Flora of Central Texas.* Fort Worth: Prestige Press. 320 pp.

Richardson, A. 1990. *Plants of the Rio Grande Delta.* Austin: University of Texas Press. 332 pp.

————. 2002. *Wildflowers and Other Plants of Texas Beaches and Islands.* Austin: University of Texas Press. 247 pp.

Rogers, J. 2008. Clearing up some tall fescue misconceptions. Samuel Roberts Nobel Foundation, http://www.noble.org/index.html, accessed 18 March 2008.

Rollo, J. D., and E. G. Bolen. 1969. Ecological relationships of blue and green-winged teal on the high plains of Texas in early fall. Southwestern Naturalist 14:171–188.

Rosene, W., and J. D. Freeman. 1988. *A Guide to and Culture of Flowering Plants and Their Seed Important to Bobwhite Quail.* Augusta: Morris Communications Corporation. 170 pp.

Russell, D. N. 1971. Food habits of the starling in eastern Texas. Condor 73:369–372.

Sands, J. P. 2007. Impacts of invasive exotic grasses on northern bobwhite habitat use and selection in south Texas. M.S. Thesis, Kingsville: Texas A&M University–Kingsville. 106 pp.

Schacht, S. J., T. C. Tacha, and G. L. Waggerman. 1995. Bioenergetics of white-winged reproduction in the lower Rio Grande valley of Texas. Wildlife Monographs 129:3–31.

Scifres, C. J., and W. T. Hamilton. 1993. *Prescribed Burning for Brushland*

Management: The South Texas Example. College Station: Texas A&M University Press. 246 pp.

Scribner, K. T., and L. J. Krysl. 1982. Summer foods of the Audubons cottontail (*Sylvilagus auduboni*: Leporidae) on Texas panhandle playa basins. Southwestern Naturalist 27:460–463.

Sheeley, D. G., and L. M. Smith. 1989. Tests of diet and condition bias in hunter-killed northern pintails. Journal of Wildlife Management 53:765–769.

Sheffield, W. J. 1983. Food habits of nilgai antelope in Texas. Journal of Range Management 36:316–322.

Shields, R. H., and E. L. Benham. 1969. Farm crops as food supplements for whooping cranes. Journal of Wildlife Management 33:811–817.

Short, H. L. 1971. Forage digestibility and diet of deer on southern upland range. Journal of Wildlife Management 35:698–706.

Simpson, B. J. 1988. *A Field Guide to Texas Trees.* Lanham, Maryland: Lone Star Books. 372 pp.

Taylor, R. B., J. Rutledge, and J. G. Herrera. 1999. *A Field Guide to Common South Texas Shrubs.* Austin: Texas Parks and Wildlife Press. 106 pp.

Taylor, W. P. 1954. Food habits and notes on life history of the ring-tailed cat in Texas. Journal of Mammalogy 35:55–63.

Texas Commission on Environmental Quality. 2009. Texas General Land Office, http://www.glo.state.tx.us.gisdata/metadata/Counties.htm, accessed 10 May 2009.

Texas Invasive Plant Conference Proceedings. 2007. http://www.texasinvasives. org/ conference07/conference_assets/ TIPC_Program.pdf, accessed 17 March 2008.

Texas Parks and Wildlife Department. 2007. Bobwhite quail in the Post Oak Savannah and Blackland Prairie, http://www.tpwd.state.tx.us/landwater/land/habitats/post_oak/upland_ game/bobwhite, accessed 27 January 2007.

Tix, D. 2000. *Cenchrus ciliaris* invasion and control in southwestern U.S. grasslands and shrublands. Restoration and Reclamation Review, volume 6.1, http://horticulture.cfans.umn.edu/vd/h5015/rrr.htm, accessed 18 March 2008.

Tjemeland, A. D., T. E. Fulbright, and J. Lloyd-Reilley. 2008. Evaluation of herbicides for restoring native grasses in buffelgrass-dominated grasslands. Restoration Ecology 1:263–269.

Tull, D., and G. O. Miller. 1991. *A Field Guide to Wildflowers, Trees, and Shrubs of Texas.* Houston: Gulf Publishing Company. 344 pp.

Turner, B. L., H. Nichols, G. Denny, and O. Doron. 2003. *Atlas of the Vascular Plants of Texas.* Fort Worth: Brit Press. 648 pp.

United States Department of Agriculture, Natural Resources Conservation Service. 2006. Plant fact sheet: bermudagrass (*Cynodon dactylon* [L.] Pers.), http://plants.usda.gov/factsheet/pdf/fs_cyda.pdf, accessed 27 January 2007.

USDA, NRCS. 2008. The PLANTS Database, http://plants.usda.gov, 11 June 2008. Baton Rouge: National Plant Data Center, accessed 8 February 2007.

Vines, R. A. 1984. *Trees of Central Texas: A Field Guide*. Austin: University of Texas Press. 405 pp.

Wagner, F. H. 1949. Notes on the winter feeding habits of north Texas bobwhites. Field and Laboratory 17:90–98.

Warren, R. J., and L. J. Krysl. 1983. White-tailed deer food habits and nutritional status as affected by grazing and deer-harvest management. Journal of Range Management 36:104–109.

Whitford, W. G. 1997. Desertification and animal biodiversity in the desert grasslands of North America. Journal of Arid Environments 37:700–720.

Williams, D. G., and Z. Baruch. 2000. African grass invasion in the Americas: ecosystem consequences and role of ecophysiology. Biological Invasions 2:123–140.

Williams, L. R., and G. N. Cameron. 1986. Food habits of Attwater's pocket gopher, *Geomys attwateri*. Journal of Mammalogy 67:489–496.

Wood, J. E. 1954. Food habits of furbearers of the upland post oak region in Texas. Journal of Mammalogy 35:406–415.

Wood, K. N. 1985. Bobwhite foods and nutrition in the Rio Grande plain of Texas. M.S. Thesis, Kingsville: Texas A&I University. 60pp.

Wood, K. N., F. S. Guthery, and N. E. Koerth. 1986. Spring-summer nutrition and condition of northern bobwhites in south Texas. Journal of Wildlife Management 50:84–88.

Wrede, J. 2005. *Trees, Shrubs, and Vines of the Texas Hill Country*. College Station: Texas A&M University Press. 246 pp.

Index

Page numbers in italics refer to illustrations.